THE MOST PERFECT THING

THE MOST PERFECT THING

Inside (and Outside) a Bird's Egg

Tim Birkhead

BLOOMSBURY

LONDON · OXFORD · NEW YORK · NEW DELHI · SYDNEY

Bloomsbury Publishing
An imprint of Bloomsbury Publishing Plc

50 Bedford Square
London
WC1B 3DP
UK

1385 Broadway
New York
NY 10018
USA

www.bloomsbury.com

BLOOMSBURY and the Diana logo are trademarks of Bloomsbury Publishing Plc

First published in Great Britain 2016

British Library Cataloguing-in-Publication Data
A catalogue record for this book is available from the British Library.

ISBN: HB: 978-1-4088-5125-8
ePub: 978-1-4088-5126-5

2 4 6 8 10 9 7 5 3 1

Typeset by Newgen Knowledge Works (P) Ltd., Chennai, India
Printed and bound in Great Britain by CPI Group (UK) Ltd, Croydon CR0 4YY

MIX
Paper from
responsible sources
FSC® C020471

To find out more about our authors and books visit www.bloomsbury.
com. Here you will find extracts, author interviews, details of forthcoming
events and the option to sign up for our newsletters.

For my mother and for Erick Greene

Contents

Preface

This book has had a long incubation but was spun into life by a chance encounter. I was watching a wildlife programme on television one evening in 2012 when a well-known presenter appeared beside a cabinet of eggs in a museum. Opening one of the drawers, he took out an egg. It was white, I remember, and he held it to the camera to demonstrate its size and unusual pointed shape. This is a guillemot's egg, he announced, and the reason – he said – for its extraordinary shape is so that it will spin on its axis and not roll off the narrow cliff ledge on which guillemots breed. To demonstrate, he placed the egg on the surface of the cabinet and spun it, and, sure enough, the egg rotated on the spot like a horizontal top.

I could not believe what I had just seen. Not because it is amazing, but because I was aghast that someone so revered for his natural history knowledge could make such a mistake. The spinning-on-the-spot story for guillemot eggs was debunked over a century ago, and here it was being given new life to an audience of millions.

You can spin a guillemot egg on its long axis, especially if, like the television presenter, you are using a blown (empty) museum egg. But that is not how a real egg – full of either yolk and albumen or a developing embryo – operates.

When I wrote to the presenter pointing out that what he had told us was wrong his initial response was understandably grumpy. I offered to send him the scientific papers on this topic so he could read the relevant research. I was about to post them when I had a sudden crisis of confidence. Here I was telling a television star how to behave when I could be wrong myself. I decided to reread the papers.

I have studied guillemots continuously since the early 1970s, in England, Wales, Scotland, Newfoundland, Labrador and the Canadian High Arctic. I've lived and breathed guillemots for forty years and I've read pretty well everything that has been written about them. I last looked at the articles dealing with guillemot egg shape twenty years previously, which is why I suddenly questioned my memory and decided to reread them. It was just as well I did, for the data and conclusions were both opaque and much messier than I remembered. I was also shocked by the realisation that most of those scientific papers on egg shape in guillemots were written in German. Some of them had an English summary, but, as every scientist knows, the summary, or abstract, while purporting to provide an accurate synopsis of the paper's content, is often a shop window in which authors make their results seem stronger than they really are.

The first paper, and the source of the conventional wisdom, was that the guillemot's pointed egg shape allowed it, not to spin like a top, but to roll in an arc, and it was rolling-in-an-arc that was said to prevent it from falling.

As I read the paper's summary and looked at the graphs and tables – using a German dictionary to translate their captions – I felt that something wasn't right. It didn't add up. I found a German-speaking student in my department and paid him (quite a lot) to translate the paper word for word. The study's conclusions were far from clear, and, as we'll see, even rolling-in-an-arc was not especially convincing.

I decided to reinvestigate. Although this was an old problem, re-entering the world of guillemot eggs was like exploring a new world, with paths leading in all sorts of new directions in what has

since been an exhilarating journey. At one level it sounds trivial –
who cares why guillemots lay pointed eggs? But at another it was
wonderful, encompassing everything that science is supposed to
be. And I say that with due modesty. Much of science has been
distorted by government-imposed assessment exercises that result
in short-term financially motivated studies whose results are often
overstated or occasionally even falsified. My egg project had an air
of adventure about it and to my mind that's what science should be:
an adventure.

One of the first things I discovered was that the guillemot's
egg – and unless I say otherwise, guillemot means the common
guillemot* – was the most popular and sought after for collections.
In the past when egg collecting was widespread, without several
guillemot eggs your collection was incomplete. Why? Because guillemot eggs are seductively beautiful – large, brightly and infinitely
variable in colour and pattern, and, as seen on TV, very oddly
shaped.

My guillemot research started in 1972 on Skomer Island off the
western tip of South Wales and I have been back each year since
then. Bounded by 200-foot basalt cliffs, Skomer is one of Britain's
most important seabird colonies. It enjoys complete protection
today but in the past – like almost every other British seabird
colony – Skomer's cliffs were plundered for their eggs.

In May 1896 Robert Drane, founding member of the Cardiff
Naturalists' Society, accompanied Joshua James (J. J.) Neale and his
wife and ten children on a trip to Skomer. There are two accounts

---

* There are two species of guillemots in the genus *Uria*: the common guillemot, *Uria aalge*,
which breeds around the UK, and Brünnich's guillemot, *U. lomvia*, which breeds further
north. Both species breed in the Atlantic and Pacific. In North America the two species are
referred to as the common murre (rhymes with 'fur') and thick-billed murre, respectively.
There are other more distantly related species, including the black guillemot, whose breeding
biology is rather different. The scientific names of birds mentioned in the text are given on
pages 270–4.

of their visit: Neale's, which is perfectly normal, and Drane's, which is somewhat surreal. Drane entitled his *A Pilgrimage to Golgotha* but did not reveal where or what Golgotha was, 'to avoid what we have found to be the detrimental effect to natural history of giving unreserved publicity to the places where the following notes were made'. Golgotha is an allusion to Calvary, but literally means the place of skulls, because then as now Skomer is littered with the skulls of seabirds, mainly Manx shearwaters killed by predatory great black-backed gulls. Although the number of skulls and corpses seemed (and still seems) incredible, so too is the breeding population of the nocturnal shearwaters, with current estimates at over 200,000 pairs.

During their visit, two of Neale's sons scrambled around the cliffs collecting guillemot and razorbill eggs for Drane. And not without risk. Neale laconically describes how at one point as his eldest boy climbed up from the boat to a group of guillemots, the rock he was holding came away in his hand and he fell backwards, hit the cliff and somersaulted into the sea. It isn't clear how far he fell, but luckily the impact wasn't enough to render him unconscious and he swam to a rock. His brother was close by and managed to haul him into the boat. A day or so later he had recovered from his fall, and as his father commented, 'it cured him of climbing . . .'

Neale doesn't say where this incident occurred, but there aren't that many places you can step from a small boat and climb up Skomer's cliffs. My guess is that it must have been at the east end of the island at a place known as Shag Hole Bay where there are still plenty of guillemots, even though the shags that once nested there have gone.

Before that accident, Neale's boys had collected a considerable number of eggs that were subsequently blown and added to the many hundreds at Cardiff's natural history museum, which had been accumulated by 'friends who have annually visited the coasts of South Wales'. Drane celebrated the beauty and variety of these eggs in a paper published in his *Transactions of the Cardiff Naturalists' Society*. He selected thirty-six guillemot and twenty-eight razorbill eggs taken from Skomer's cliffs and arranged the lithographs four to a page, to illustrate the extraordinary range of colours, patterns, shapes and sizes in a way no other publication has done previously or since. The images are superb but Drane's accompanying text describing the visit to Skomer and the eggs is execrable.[1]

Drane was certainly not the first to collect guillemot eggs from Skomer. There is a photograph dating from the late 1800s of the owner Vaughan Palmer Davies's daughters and friend – all in their twenties – blowing guillemots' eggs, either for themselves or to give away as souvenirs.[2]

What ends up in private or museum collections are the eggs' shells – the lifeless, outermost cover of the bird's egg. The remainder, the egg's contents that could have contributed to the creation of new life, are either eaten or discarded. Most of us have in our minds two contrasting images of birds' eggs. The first is of beautiful, often intricately coloured shells from a diversity of bird species portrayed in books or in museum cabinets. The second is the ubiquitous hen's egg either entire in its plastic carton, or as yellow yolk surrounded by its transparent 'white' in a kitchen bowl.

But there is much more to birds' eggs than either of these images imply. I have spent a lifetime – four wonderful decades – studying a wide variety of birds and their eggs, and my aim is to take you on a journey like no other. It is a voyage into the secret world of the bird's egg; it is territory that few have trodden before, and none taking the route I have planned. We will travel from the outside of the egg towards its genetic centre and on the way we will witness the three great events of bird reproduction. In doing so we will see the bird's egg for what it really is – an independent, self-contained embryo-development system.

After contemplating the irresistible appeal of eggs in Chapter 1, we look next at the most obvious part of an egg, its shell, examining how it is created (Chapter 2), how it acquires its wonderful shape (Chapter 3), its often beautiful colours (Chapter 4), and what those pigments and patterns mean in the bird's life – asking why they have evolved (Chapter 5). Moving inwards from the shell we next encounter the egg white, or albumen, which in its fresh, glutinous state we rarely give much thought to, but which turns out to be more sophisticated and crucial in safeguarding the embryo's development than you might ever have imagined (Chapter 6). As we continue inwards we reach the yolk (Chapter 7). This is the ovum proper, equivalent to a human egg, except that in birds it is huge because it is full of food – the liquid yolk itself – for the growing embryo. The female's genetic material lies as a tiny pale speck on the yolk's surface and if it is lucky enough to meet up with a male's genetic material in one (or more) sperm, it will (with luck) result in an embryo. Our outside-to-inside journey isn't a direct one, and we will occasionally need to make a short detour to a mirador that allows us to take in the entire view of where we've been and where are going. When talking about the yolk, for example, I pause to tell you how it is created in the bird's ovary. The climax to all this, one might imagine, is fertilisation – the moment at which the female's genetic material melds with that of the male – but in fact fertilisation is the first of three major events in the life of an egg. The other two are laying, and the egg's eventual hatching some ten to eighty days later, depending on the species, to produce a chick (Chapter 8).

Think of this book as your travel guide for a journey like no other. As with most guidebooks mine contains a road map – in this case, showing the layout of the female bird's reproductive tract (page 24).

It's pretty straightforward; essentially a motorway with an entrance and exit, several distinct regions, but no turn-offs, and you may want to refer to this occasionally to locate yourself. My guide also contains three key depictions of some architecture: two showing the way the egg's contents are arranged (pages 25 and 130), and a third on page 43 illustrating the eggshell's ingenious construction. There are some other images too, but those three are the most important.

There's a vast literature on birds' eggs, mainly because the poultry industry has invested millions of pounds in producing the perfect egg. Perfect for the market – not necessarily for chickens. Pretty well everything we know about birds' eggs originates from research first conducted by poultry biologists. And what a great job they've done. Motivated, in part at least, by commercial success, this is robust science that is often conducted on a scale that other biologists can only dream about. However, before we set off and see what's been discovered, it is important that you know that I, and the scientific community at large, don't have all the answers. Despite all this research, because most of it has been conducted on a single species there's still a great deal we don't know. In the current economic climate there's a tendency for researchers to justify their existence by overegging their results and exaggerating what they know. For me, knowing what we don't know is extremely important – it is what makes research exciting – and I make no apologies for highlighting the gaps in our knowledge. I do so in the hope that this will encourage other researchers to tackle some of the outstanding questions.

What I have tried to do is to bring together all those aspects of eggs that I think are interesting from a biological point of view. I have also tried to provide a sense of how and when key discoveries were made. Chicken eggs have been a ubiquitous part of human history and we hardly give them a second thought, rarely stopping to think about how they are constructed, or what the different parts do; and, of course, because the eggs we buy from the supermarket are unfertilised and unincubated, we see only a fraction of the biological miracle that eggs are. Our familiarity with the eggs of one species has blinded us to the extraordinary diversity of egg size, shape and structure across the ten thousand other species of birds that currently exist in the world. My aim, in short, is to tell you what we know and to reintroduce some wonder into this everyday miracle of nature.

Writing in 1862 the American women's rights activist Thomas Wentworth Higginson said: 'I think that, if required on pain of death to name instantly the most perfect thing in the universe, I should risk my fate on a bird's egg.'[3]

And eggs *are* perfect in so many different ways. They have to be, for birds lay and incubate in such an incredible diversity of habitats and situations, from the poles to the tropics; in wet, dry, clean and microbe-infested conditions; in nests and without nests; warmed by body heat and without body heat. The shape, colour and size of eggs as well as the composition of their yolk and albumen all constitute the most extraordinary set of adaptations. The fact that birds' eggs also provided biologists with their first insights into human reproduction makes their story even more momentous.

We start, not on Skomer, but on England's east coast, at Bempton Cliffs on the Flamborough headland.

I

Climmers and Collectors

Without the knowledge of fowles natural philosophie was very maymed.
Edward Topsell, *The Fowls of Heauen or History of Birds* (1625)

The huge, sheer limestone cliffs gleam with a startling white-ness in the bright sunlight. Following the sharp edge of the land towards the east you can see the Flamborough headland; to the north lies the holiday town of Filey, and out of sight to the south is Bridlington, another resort. Here on the Bempton cliff top, however, Filey and Bridlington might as well be a hundred miles away, for this is a wild place: benign in the sunshine, but awful on a wet and windy day. On this early summer morning, however, the sun is shining; skylarks and corn buntings are in full song and the cliff tops are ablaze with red campion. The path along the cliff top traces the meandering line that marks the fragile farmland edge and where at each successive promontory a cacophony of sound and smell belches up from below. Out over the cobalt sea birds wheel and soar in uncountable numbers and there are many more in elongated flotillas resting upon the water.

Peering over the edge you see there are thousands upon thousands of birds apparently glued to the precipitous cliffs. The most conspicuous are the guillemots packed tightly together in long dark lines. En masse they appear almost black, but individually in the sun these foot-tall, penguin-like birds are milk-chocolate brown on the head and back, and white underneath. Their dense, velvety head feathers and dark eyes suggest a wonderful gentleness, and they are gentle for much of the time, but if roused they can use their long pointed bill to great effect. Above and below the guillemots are pristine white kittiwakes, kitty-waking and squealing from their excrement-encrusted grassy nests. Less numerous and often concealed in crevices are the razorbills, known locally as tinkers for their sooty dorsal plumage. And, more sparsely still, there are the sea parrots or puffins with their glowing red beaks and feet, and which, like the razorbills, nest out of sight among the limestone fractures. The soundscape is a mix of squeaky soprano kittiwakes overlaying a tenor chorus of growling guillemots, with the occasional high-pitched hum from a contented puffin. And the smell . . . well, I love it and its associations, but – and at the risk of muddling my metaphors – it is an acquired taste.

It is June 1935 and at a point known as Staple Newk* the vista opens with the breathtaking sight of a man suspended from a 150-foot-long rope over the sea. Swinging precariously out from the sheer limestone face he glides back in towards the rock wall, stops and clings like a crab to the cliff. Watching through field glasses from a safe vantage point on the cliff top is George Lupton, a wealthy lawyer. In his mid-fifties, he is above average height, with a modest moustache, deep-set eyes and a prominent nose: his collar and tie, tweed jacket and manner all signal his affluence. Lupton watches as the man on the rope forces the guillemots to depart

* Pronounced 'Stapple Nuck', meaning a staple or pillar and a corner.

in noisy panic, abandoning their precious, pointed eggs, some of which roll away and smash on the rocks below. Most of the remaining eggs are orientated with their pointed end towards the sea. The man on the rope takes them one after another, placing them in his canvas shoulder bag that is already bulging with loot. With the ledge clear of eggs, he pushes off with his feet to swing out and back to another location slightly further along to continue his clumsy plunder. Oblivious of the climber's safety, Lupton is almost beside himself with excitement at what lies inside that canvas bag. On the cliff top three other men sit one behind the other, with the rope secured around their backs, ready when the signal comes to pull like oarsmen until the climmer emerges safely from over the cliff edge.

Yorkshire dialect has reduced these climbers to 'climmers' and past events to cliffs 'clumb'.

George Lupton has travelled by train from his home in Lancashire. He's been here for over a month and, like the other egg collectors, is staying in Bridlington.[1]

On this beautiful morning the cliff tops are busy with people and there is a holiday atmosphere. Little huddles of tourists watch in awe as the climmers descend, dangle and are hauled back up from the rock face with their bounty.

The bag is emptied and the eggs are placed in large wicker baskets. The chalky clunking of the thick-shelled eggs is music to Lupton's ears. The climmer, Henry Chandler, still in his protective policeman's helmet, smiles to himself for he knows that somewhere in his bag is a specimen Lupton desperately wants and is prepared to pay good money for. Identified as the 'Metland egg' and named for the section of cliff owned by the adjacent farm, this distinctively coloured egg – described as a 'brownish ground with a darker reddish brown zone' – has been taken each year from exactly the same spot, a few inches square, since 1911 – for over twenty consecutive years.[2]

George Lupton is obsessed by guillemot eggs. The Metland egg, although special, is one of many. The climmers have known for decades, probably centuries, that female guillemots lay an egg the same colour in precisely the same place year on year. Indeed, the climmers also know that after the first 'pull' – the season's first take of eggs – females will lay an almost identical replacement egg at the same spot a fortnight later. After that is taken they'll lay a third, and very occasionally a fourth. Lupton's lust has meant that in its twenty-year breeding life, the Metland female has never once succeeded in hatching an egg or rearing a chick. The same is true for thousands of guillemots and razorbills along these cliffs, for the climmers farm the eggs here on an industrial scale.

Men have descended Bempton's cliffs to harvest seabird eggs since at least the late 1500s. The farmers whose muddy fields led down to the cliff edge assumed ownership of the 'land' – in reality a fragile rock face – that runs vertically down to the sea below. Gangs of three or four men – comprising a climmer and three anchor-men – often several generations of the same family, work the cliffs, year after year, decade after decade.[3]

Initially the eggs were taken for human consumption. They are twice the weight of a hen's egg and are excellent scrambled. Boiled, they are slightly less appealing – at least to me – because the 'white' (the albumen) remains slightly blueish in colour and sets less hard than does that of a hen's egg. This didn't stop guillemot eggs being eaten in unimaginable numbers wherever they were available, not just at Bempton but across the coastal fringes of the entire northern hemisphere. In areas where guillemots bred on low-lying islands, as they do in North America, they were easily exploited and often to local extinction. It was too easy: guillemots breed in such dense aggregations that discovering a colony was like winning the lottery. Eventually, only those birds breeding in the most remote or inaccessible places had any chance of rearing offspring. One of the furthest flung breeding colonies – forty miles off Newfoundland's north-east coast – is Funk Island, a name reflecting the foul (or fowl?) smell emanating from hundreds of thousands of birds. Prior to the discovery of the New World, the Native American Beothuk braved the treacherous seas and paddled out to Funk Island in their canoes to feast on the eggs of guillemots and great auks and on the birds themselves. Their visits were probably infrequent enough to do little harm, but, once European seafarers discovered Funk Island and the other seabird colonies along the north shore of the St Lawrence River in the 1500s, the birds were doomed.[4]

As elsewhere, the Bempton climmers ensured that the eggs they collected were fresh simply by chucking off all the eggs they found on their first visit, and then returning every few days throughout the season to remove the new ones as they appeared. Estimates of the number of eggs taken each year at Bempton are hopelessly variable. Some say more than 100,000, others a few thousand. It was certainly thousands and the best estimates from the 1920s and 1930s, when Lupton was collecting, are annual totals of about 48,000. There were once a lot of guillemots at Bempton, but as egging continued the number of birds inevitably decreased. The decline was accelerated by the creation of the railways – to Bridlington in 1846 and to the village of Bempton itself a year later – providing easy access to those from London and other urban centres seeking the cheap thrill of shooting seabirds. Shooting not only killed and maimed hundreds of birds – mainly guillemots and kittiwakes – but each shot flushed incubating birds from their ledges causing a cascade of eggs on to the rocks or into the sea below.[5]

Lupton was one of several collectors in cahoots with the Bempton climmers. It was a lucrative arrangement for those who risked possible death on the end of a rope since they quickly came to recognise the gleam in the collector's eye and their insatiable passion for particular eggs. Possession was everything and while the collectors bartered with the climmers they also had to compete with each other. The climming gangs got on well by respecting each other's territorial boundaries, but competition between individual collectors was often intense. One was said to have pulled a gun on another in an argument over a particularly desirable egg.[6]

Sam Robson, born in 1912 and one of the climmers who supplied Lupton with eggs, recounted, with wonderful Yorkshire enunciation, what it was like:

> You went by colour a lot, for collectors' eggs: if you saw an unusually marked one, you'd take care o' that, and wait 'till these collectors came. In them days, eggs was same as coin-collecting or summat: they'd get the set, and they used to trade 'em or flog 'em. They used to come all together did collectors: you'd get as high as four or five staying in the village. It was their profession to collect eggs, and sell 'em: a lot of 'em was dealers for other collectors . . . So it was more or less like an auction at the cliff top, sometimes . . . It was a gamble, what they would pay: you demanded so much and they'd barter you if they could, to beat you down. We took what we could get, because we wanted rid on 'em: we didn't want eggs, we wanted money.[7]

The scale of the climmers' and collectors' activities is all too apparent if you check the catalogues or visit the egg collections of different European and North American museums. Almost without exception, each museum has more eggs from Bempton than almost any other location, including those in their own country. Even the modest teaching museum that I curate in Sheffield has two trays of guillemot eggs dating back to the 1830s, most of which have scrawled on them in semi-legible pencil Bempton, Buckton, Filey,

Scarborough and Speeton – all names of locations where eggs were obtained from the Flamborough headland.

I'm Yorkshire born and bred, and when Skomer Island was inaccessible during the winter months of my PhD years, I came to Bempton to see what guillemots get up to during their mysterious out-of-season visits. Leaving my parents' home near Leeds at 3 a.m., I drove through the dark, arriving at the cliffs as it was getting light, just before the guillemots began to fly in from the sea. They would appear very suddenly en masse in the half-light and their noisy ensemble sounded like a celebration, and that's what it was: a raucous, rapturous meeting of partners and neighbours – guillemots reunited.

It was always unbelievably cold on these visits, usually with a strong wind whipping in off the North Sea forcing me to huddle below the cliff top in a pathetic attempt to retain some body heat. Notebook in hand and peering through my retina-wrecking Hertel & Reuss telescope, I kept notes on the birds' activities, thrilled by what was – and still is – an extraordinary wildlife spectacle. In contrast to the birds, for me this was an intensely solitary experience since at that time there was no reserve building, no car park and no people – especially in midwinter. I have an immense affinity for Bempton and indeed for the entire Flamborough headland whose history oozes and drips into my imagination like the guillemot guano from the cliffs themselves. I particularly like the fact that between them the climmers and collectors – amateur ornithologists – created the scientific foundations of guillemot biology.

In Lupton's day climming was a tourist spectacle and you could buy postcards in the nearby resorts showing the climmers dangling on the end of a rope or with basketfuls of eggs on the cliff top, with captions like 'a good bag' or 'a good pull'. Climming was also a business catering to a varied clientele, from the casual visitor who merely

wanted a guillemot egg as a souvenir, to the more daring tourist –
mostly women, apparently – who went over the edge to collect an
egg for themselves, to the fanatical collectors like Lupton patrolling
like a predator along the cliff top and waiting impatiently for the
climmers to produce unusual specimens. Lupton even allowed his
eleven-year-old daughter Patricia to be lowered over the edge to
collect eggs for herself.[8]

Guillemot eggs are extraordinary in many different ways, but
especially in size, colour and pattern. Most early writers said that
no two were alike and it was this seemingly infinite variety of colour
that mesmerised George Lupton. He was not alone. Dozens of
collectors were greedy for guillemot eggs, but Lupton was unusual
in being the only collector, or oologist, as they called themselves,
to focus almost his entire nervous energy, and the contents of his
wallet, on the guillemot's ovarian output. Another Bempton collector, George Rickaby of Nottingham, in 1934 described Lupton's
collection of over a thousand unusual guillemot eggs as being 'the
world's best'.[9]

The 1930s, when Lupton, Rickaby and others were active along
the Bempton cliff tops, was the heyday of egg collecting in Britain.
We look back on those times with both wonder and dismay. Once
deemed a harmless part of every country boy's childhood that
occasionally swelled into an adult preoccupation, egg collecting
is now unacceptable and illegal. The irony is that in the past egg
collecting was just one of many ways of connecting with nature.
For individuals like Lupton who failed to outgrow their juvenile
pursuit, egg collecting became an obsession. He sold his collection
of guillemot eggs a decade before the Protection of Birds Act of
1954 made criminals out of those who had previously been mere
eccentrics.[10]

Collecting birds' eggs began in the 1600s when physicians,
savants and others interested in the natural world started to acquire

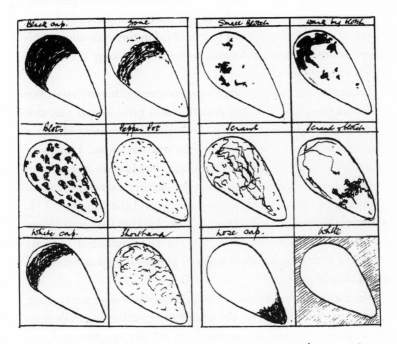

The names used by the Bempton climbers to categorise the patterning on guillemot eggs. Top row (from left to right): black cap, zone, small blotch, dark big blotch; middle row: blots, pepper pot, scrawl, scrawl and blotch; bottom row: white cap, shorthand, nose cap, and white. From George Rickaby's diary: Whittaker, 1997.

artefacts and create cabinets of curiosities. Among the first of these was the great Italian naturalist Ulisse Aldrovandi, whose museum opened in 1617. His collection contained – among many other things – an ostrich egg, amazing for its sheer size, but also several monstrously large and deformed hen's eggs. Also part of his collection was an oversized (presumably double-yolked) goose egg and an egg from a hen that had once been a cockerel.[11]

Another Renaissance man who had eggs in his cabinet was Thomas Browne, a brilliant physician based in Norwich, England. Browne's wide range of interests included the new scientific natural history, and among his many achievements was the first account of the birds of Norfolk. After visiting Browne in 1671, John Evelyn, writer, gardener and contemporary of Samuel Pepys, reported in his diary on 18 October 1671:

> Next morning I went to see Sir Thomas Browne (with whom I had sometime corresponded by letters tho never saw before) whose whole house & garden being a paradise and cabinet of rarities, & that of the best collection, especialy medails, books, plants, natural things, did exceedingly refresh me after last nights confusion: Sir Thomas had amongst other curiosities, a collection of the eggs of all foul and birds he could procure, that country (especialy the promontorys of Norfolck) being (as he said) frequented with the severall kinds, which seldome or never, go farther into the land, as cranes, storkes, eagles etc. and a variety of water-foule.[12]

Perhaps the most significant of these early naturalists with an interest in eggs was Francis Willughby, who, together with John Ray, published the first 'scientific' book on birds in 1676. It was written by Ray and entitled *The Ornithology of Francis Willughby* in honour of his friend and collaborator who died in 1672 at the tender age of thirty-six. Published first in Latin in 1676 and in English in 1678, I refer to this important book simply as *Ornithology*.

Willughby knew of Browne and they may have corresponded, but we don't know if they ever met or whether Browne encouraged Willughby to collect eggs and other natural history artefacts. But we do know that Willughby once had a cabinet of curiosities because his daughter Cassandra refers to it in a letter, after collecting her late father's belongings, saying that they included: '. . . a fine collection of valuable meddalls, and other rarities which my father had collected together of dryed birds, fish, insects, shells, seeds, minerals and plants and other rarities . . .'[13]

As I read this, I assumed that Willughby's collection of biological curiosities, like those of Aldrovandi, Thomas Browne and many others, was long gone – rotted away or simply thrown out. Imagine my utter amazement when I discovered that Francis Willughby's cabinet, including his eggs, has survived as part of the family estate.

The cabinet comprises twelve drawers, most of which contain botanical specimens, and it was while I was photographing these for a friend that I pulled open the bottom drawer. As I did so, I was dumbfounded to see birds' eggs. Like the plant specimens above them, these were held – loosely, it has to be said – within different shaped compartments. Many of the eggs were broken, and all were covered by a sticky layer of grime, reflecting the fact that the family home had once been in the heart of an English coal-mining area. Some of the eggs had the species name written on them in brown ink: *Fringilla* [chaffinch], *Corvus* [carrion crow, or rook], *Buteo* [common buzzard], *Picus viridis* [green woodpecker] and herne [grey heron].

That the eggs had survived was a miracle: that the eggs were labelled was an even greater miracle, for it allowed me to verify their authenticity. Many twentieth-century collectors, including

George Lupton, failed to mark their eggs in a way that would allow anyone other than themselves to explore their history. But many of Willughby's eggs had the name of the bird written directly on the shell and in his unmistakable hand.

I arranged for Douglas Russell, the curator of eggs at the Natural History Museum, to accompany me to examine Willughby's cabinet and offer his expert opinion. Like me, he was amazed by the collection. Most of the eggs were frighteningly fragile, and even those of the larger species had aged to such an extent that, notwithstanding the grime, they were virtually transparent. Douglas was quickly convinced of the collection's originality, and hence its historical value. He told me that there is no other collection in the world as old. Prior to this moment he said, the oldest known egg in any scientific collection was probably that of a great auk, once belonging to the great Italian priest-cum-scientist Lazzaro Spallanzani and dating back to 1760. Willughby's eggs are a century older.

Private cabinets of curiosities metamorphosed into public museums in the 1800s, fuelling a passion for collecting birds' eggs. Acquisition, in the name of national pride, meant that eggs and birds (in the form of study skins and skeletons) were accumulated on an unimaginable scale. From this period onwards, the science of ornithology – conducted mainly by wealthy amateurs – became synonymous with both museums and collecting.

The same was true of other types of natural history specimens, and the acquisition of butterflies and birds' eggs had much in common. Both types of collector were driven partly by aesthetics but also by the idea of capturing the full swathe of variation that existed within a particular species: 'a passion for beauty and a lust for curiosities'. There were even some butterfly collectors who, like Lupton, focused almost exclusively on a single species. And there were others, again like Lupton, whose motivation was to use their lifeless biological trophies to create a visual spectacle without even a nod to the demands for data. The huge collections of butterflies, many still in private hands, others in provincial and national museums, are testament to that insatiable quest. It is curious, though, that despite the numbers of specimens taken the collectors of butterflies have not been demonised in the same way as egg collectors.[14]

Alfred Newton, founding member of the British Ornithologists' Union (BOU) in the 1850s, and a collector, extolled the virtues of egg collecting in his typical long-winded Victorian way: 'the fascination with this boyish pursuit has maintained its full force even in old age – a fact not so much to be wondered at when it is considered that hardly any branch of the practical study of Natural History brings the enquirer so closely into contact with many of its secrets.'[15] As Newton points out, the collecting of eggs by boys (it was never girls) was an essential part of the study of natural history. Well-known naturalists and conservationists of the twentieth century, including David Attenborough, Bill Oddie and Mark Cocker, all admitted to collecting eggs when they were young, but this merely emphasises just how important it was to their later careers.[16]

Some of the original justification for collecting birds' eggs was that, together with the skins or skeletons of birds, they would provide material from which the natural order of birds could be deduced. Indeed, identifying God's great plan was the main goal of scientific ornithology, epitomised by Willughby and Ray's *Ornithology*. The same was true for the whole of biology – zoology and botany both – in what ways are different species related to each other? It was obvious that a pattern existed: greenfinches and goldfinches are more similar to each other than they are to either dotterel or dabchicks, but the basis for those relationships was often elusive. At that time the only clues for visualising the arrangement of birds were their external and internal features: plumage colour and pattern on the one hand, gut, cranium or syrinx on the other. But eggs: their colour, shape and structure – it was thought – might also contribute to this scientific endeavour.

Had God not been hell-bent on moving in mysterious ways and creating interesting intellectual challenges for his followers, the pattern might have been obvious. But, of course, the arrangement of birds is not a product of God's wisdom. It is the result of millions of years of evolution, which also works in ways that often seem mysterious. Evolutionary processes are particularly striking in the manner in which they sometimes create similar structures in unrelated species. Hummingbirds in the New World and sunbirds in the Old World both feed on nectar which they extract from flowers using their long tongue and bill and they both have iridescent plumage. Despite their physical similarity hummingbirds and sunbirds do not have an immediate common ancestor – they have evolved completely independently. Similar environments create similar selection pressures and result in similar body forms through a process known as 'convergent evolution'. Long after the idea of God's wisdom in the form of natural theology had been replaced by Darwin's natural selection, convergent evolution continued to be not only the best evidence for natural selection, but also the bugbear of those trying to understand the relationships between birds. It wasn't until molecular methods – the examination of genetic signatures – provided a truly objective way of doing so in the early twenty-first century that scientists finally felt they had a reasonable picture of the evolutionary history of birds and the relationships between them.[17]

Throughout the entire 400-year period that ornithologists struggled to understand the relationships between different groups of birds, museum specimens were vital. Skins and skeletons were crucial for at least it was possible to see some patterns there. But in this respect eggs proved to be almost entirely useless. Realising this late in life, Alfred Newton wrote: 'I must confess a certain amount of disappointment as to the benefits it was expected to confer on systematic ornithology . . . Oology taken alone proves to be a guide as misleading as any other arbitrary character.'[18] It soon became increasingly difficult to justify the continued collecting of eggs on scientific grounds.

There is something sensual about eggs. Of course there is: they are part of sexual reproduction, but birds' eggs have an erotic aura all of their own. Perhaps their wonderful curves trigger deep-rooted visual and tactile sensations among men. As though to confirm this, one book I found on egg collecting drew parallels between eggs and the female form, illustrated in a series of seductive ovals and spheres.[19] This may also be one reason Fabergé's eggs are so popular: an expensive nuptial gift that fuses sensuality of form with the ultimate symbol of fertility.

More than a hint of something sexual is apparent in Philip Manson-Bahr's recollection of Alfred Newton: 'Although a confirmed misogynist he could be charmingly polite to the opposite sex, but he held firmly to his principles that his museum and its treasures were not for feminine eyes and he would never vouchsafe them even so much as a peep at his egg collection . . . To watch Newton ogling his eggs was another cameo. He adored them.'[20]

There's an additional reason for thinking that it is the three-dimensional shape or form of eggs that encapsulates their beauty. Compared with the birds that feature so extensively in works of art, paintings of eggs are extremely scarce, suggesting that rendering them in two dimensions simply doesn't do it for most people. In contrast, egg-shaped sculptures – like those of Barbara Hepworth and Henry Moore, for example – have huge appeal.[21]

On a cold winter's day early in 2014 I visited the ornithology department at the Natural History Museum in Tring, Hertfordshire, to look at Lupton's assemblage of over one thousand Bempton eggs. Since most egg collectors kept records of where and when they obtained their specimens, I naively assumed Lupton would have done the same. Far from it! Lupton seems to have relied largely on his memory to tell him where and when he had acquired each specimen. In a few cases there are uninformative notes on scraps of paper lying alongside the eggs, but what they mean is anybody's guess. To put it bluntly, Lupton's collection is a shambles, but this is what it was like when the museum acquired it.[22]

As a scientist it almost made me weep: so much information so carelessly lost! Perhaps data cards were irrelevant to someone like Lupton whose focus was aesthetic rather than scientific. Some of his cabinets in Tring are indeed beautiful. There are trays of almost identical twos, threes and fours, apparently from the same female guillemots in the same year; or from the same female in different years. Another tray holds thirty-nine extraordinarily scarce completely white unmarked eggs, which Lupton says, in a carelessly scrawled note, came from three different females all on the same Bempton ledge! Another box contains twenty unusually coloured eggs: white ground overwritten in what looks like red Pitman shorthand, but this time – improbably – from widely separated locations around the British coast, contradicting the belief that no two guillemot eggs are alike.

It is with a mixture of disappointment and awe that I look at Lupton's eggs. I'm disappointed because, with no data cards, this vast array of wonderful eggs has almost no scientific value. On the other hand, I'm amazed by the sheer diversity of eggs, by the scale and nature of his obsession and by his artistic inventiveness. When I express my frustration about the lack of data to Douglas, the curator, he responds by asking whether my glass is half empty or half full, for without Lupton's collection, he says, there'd be nothing to write or think about. My glass is half full. More than half full in fact, because I can see Lupton's aesthetic enjoyment of his eggs. I can also see how fortunate it is that no one has bothered to curate Lupton's collection for this would have inevitably destroyed their beautiful arrangements.

If someone later discovers Lupton's data cards, then possibly, just possibly, we will be able to match them to the eggs and start to explore how much year-to-year variation there was in egg size; how similar the colour and shape remains across different eggs laid by the same female within a season, or, as in the case of the Metland eggs, across much of a female's lifetime.[23] There are many ways we could gently interrogate Lupton's eggs. It may still be possible.

I suspect there *are* no cards and no master data sheet that would allow us to crack the code. Everything I can see of Lupton's collection exudes aesthetics and excludes science. Perhaps most telling are torn scraps of light-green paper with almost illegible pencil scribbles placed inside the cabinet drawers – many of them signed with his initials. The messages on these pieces of paper, such as 'x4' or 'x3', are as brief as they are cryptic; and why would anyone place *signed* notes in their trays of eggs if they also had data cards or a master sheet?

Lupton's glass-topped egg trays, measuring two foot by two foot (60 x 60cm), now lie within the white plastic drawers of the British

Museum's cabinets. At Douglas's suggestion, we got them out – all thirty-seven of them – and laid them against each other on tables, benches and the floor. Only then was the full visual impact apparent. Lupton must have spent months sorting his eggs, deciding on different arrangements, and then seeking those eggs that would make the arrangement complete, for the whole purpose of this was display. Lupton's egg trays were the individual feathers of a peacock's train; each egg an eye-spot on the peacock's tail, and a display just as daring, just as spectacular and every bit as difficult to interpret.

Lupton's collection is organised according to almost every conceivable oological criterion one could think of: colour, size, shape and texture. But these terms don't do the display justice, for colour encompasses ground colour, type and hue of markings and the distribution of marks across the surface of the egg. One of his most subtle arrangements is of twelve groups of four horizontally arranged eggs, all of which have fine pepper and salt speckling on different coloured grounds; pale blue, pale green, yellow ochre and white, and with adjacent groups arranged as mirror images of each other. It is art.

In another drawer there is something more extraordinary still: pairs comprising one guillemot and one razorbill egg, obviously different in their shape (the razorbill's egg is much less pointed), but absolutely identical in colour and patterning. This is remarkable. Razorbills, which typically breed among guillemots, but always in isolation and in crevices, produce much less variable, less colourful eggs than guillemots. If asked, I reckon I could correctly identify well over 90 per cent of eggs to razorbills or guillemots based simply on their colour and their pattern. But Lupton found that minuscule proportion of both species whose eggs resembled the other species, making me wonder what proportion of the egg-colour genes the two species have in common.

I tried to imagine Lupton building up these arrangements year on year. Pacing the Bempton cliff tops, seeing a haul of climmers' eggs, and knowing which ones he needed for his collection. Over the winter months he must have spent long hours examining his eggs to know exactly what he'd got, and what he needed. I can also imagine him reliving the moments of breathless excitement as he acquired particular eggs whose addition brought the collection a step closer to perfection.

Lupton is history; egg collecting is largely history, too. Although its scientific worth may be limited, collections like Lupton's still have a value. Several museum curators I spoke to while I was writing this book told me how in the past they had destroyed hundreds of guillemot eggs – mostly from Bempton – because they had no accompanying data, and were therefore 'scientifically worthless'. It made me cringe. So often, something once thought of as useless is – seen in a different light or with different technology – actually valuable. In ways that could not have been imagined by the original collectors, it is now possible to take a minute fragment of eggshell and extract the DNA from it, allowing us to identify the genotype of the female that laid it.[24] And who knows: as molecular methods continue to develop it might even be possible to reconstruct Lupton's missing data from the genetic notation embedded within his eggs.

Collections like Lupton's also have a cultural value. I'm acutely aware as a scientist that it is all too easy to view the world through a single lens – a lens scientists often assume to be superior. The artistic arrangements of Lupton's guillemot eggs are unique; they are worthy of an exhibition in themselves, and I can easily see that if taken from the dark confines of Tring's Natural History Museum they would inspire artists and others to see the natural world in a different light.

Not only would they see the aesthetic perfection of eggs; they might also ask about their biological perfection: how they are constructed and how the seemingly endless variations in colour are applied; why eggs vary so much in shape and size; why their yolk and albumen, which superficially seem so uniform, are in reality so variable; how that single female cell is fired into life by one or more sperm; and how after a matter of a few weeks a new life breaks free from the fragile yet robust structure we call the shell.

And the shell is where we start our exploration and make our way from the outside to the inside of the bird's egg.

2

Making Shells

Biologists can learn many things about bird species from their eggs.
R. Purcell, L. S. Hall and R. Coardso, *Egg & Nest* (2008)

The psychological distance between a museum drawer full of lifeless eggshells and the living oviducts in which they were created is vast. In many cases it is geographically vast, too. Few who view eggs in a museum ever make the connection. One reason for this disconnect is that not many of us today have the opportunity to see or touch the living, breathing eggs of wild birds.

The developing embryo is protected from the outside world by the egg's hard chalky shell, but the shell also creates a connection with that world. How do you make a structure that keeps microbial aliens out, but at the same time allows the embryo inside to breathe; a shell that is strong enough to withstand the full weight of an incubating parent, but weak enough to allow the chick to eventually break free? Evolution has done a fine job of devising 'a self-contained life support system' – what is essentially an external placenta and premature baby unit.[1]

Much of what we know about the construction of eggshells comes from the pioneering studies of a nineteenth-century

German named Wilhelm von Nathusius – a man technically ingenious but biologically ignorant. Born in 1821 into a wealthy aristocratic dynasty, he studied chemistry in Paris with the intention of going into the family's porcelain factory. He preferred agriculture, however, and on taking over one of the family's estates in Magdeburg on the River Elbe, spent the rest of his working life developing novel farming methods. He published extensively on agriculture and was knighted for his efforts by the Prussian king in 1861. Eggshells were his hobby, but Nathusius's biological views were anything but mainstream, for he was one of a group of German biologists who despised Darwin and rejected the recent ground-breaking discovery by Matthias Schleiden and Theodor Schwann that cells were the basis for all life.

Nathusius's old-fashioned, contrarian opinions didn't prevent him from conducting the most detailed comparative study of birds' eggshells to date. Living far away from any university, and, one suspects, working largely in isolation, he had his own laboratory and was extraordinarily inventive in developing new methods of microscopy. Eggshells are hard, both literally and metaphorically, but, using an array of corrosive chemicals, coloured dyes and tremendous ingenuity, Nathusius found ways of exploring and documenting the eggshell structure of no fewer than sixty different bird species. His investigations included a wide range of species – ostriches, kiwis, hoopoes, wrynecks, cranes and guillemots, all from his own collections – but Nathusius's biology was curiously narrow in that he believed science should comprise little other than description. It was for this reason he rejected what he considered the unverified speculations of Darwin, Schleiden and Schwann: because their ideas weren't based on fact.[2]

In the 1960s another eggshell researcher, Cyril Tyler, a biologist based at the University of Reading in the United Kingdom, translated and made an English summary of Nathusius's thirty papers.

Tyler was simultaneously amazed at Nathusius's achievements, and frustrated by his long-winded, repetitive writing. Tyler also comments on Nathusius's complaint about how difficult it was to get his work published. This is hardly surprising given his biological tunnel vision: Nathusius was up against the much better informed (and Darwinian) Friedrich Kutter – president of the German ornithological society. To me, Nathusius is an interesting example of someone who made a major factual contribution to a particular field despite lacking any conceptual knowledge of biology – in much the same way that a surgeon can be technically brilliant without knowing anything about the way evolution created the bodies they open up and repair.[3]

Let's join the 'egg' part-way down the oviduct as it arrives at the uterus just before the shell is created, some six hours after it was released from the ovary and fertilised.

On entering the uterus (aka the shell gland) the 'egg', such as it is, is bound only by a soft membrane and is squidgy to the touch. You can recreate this stage in the egg-production process for yourself very easily by placing an egg – a hen's egg will do – in a jam jar of clear vinegar overnight.

Using a sap-green guillemot egg I'd found abandoned in a crack, I did this and watched as thousands of tiny bubbles of carbon dioxide appeared on the shell surface – the result of the acetic acid in the vinegar reacting with the calcium carbonate of the shell. The bubbles grew larger, eventually breaking free and rising to the surface of the liquid. It was like watching an Alka-Seltzer in very slow motion. After forty-eight hours the shell had disappeared completely and as I lifted the egg out from the vinegar I was slightly repulsed by its limp, wrinkled texture. The now shell-less egg lay in my hand wet and flaccid, the antithesis of an egg, but in this case still green and with some of its original dark markings. Placing the egg in a basin of water and washing away a few bits of debris, I was surprised to see

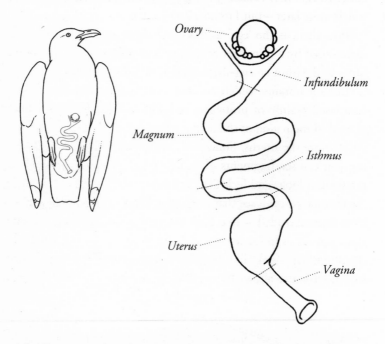

A bird's ovary and oviduct (egg tube). The regions in which different parts of the egg production processes occur are identified. This is a schematic representation – in life, the oviduct is tightly coiled upon itself.

that the leathery membrane held the egg in exactly the same shape as when it had its shell.

What the vinegar does is to reverse the process of shell formation, by eating away the calcium carbonate from the outside. Some of the vinegar may actually penetrate the shell through its pores – of which more later – but I'm not sure about that.

This, then, is the 'egg' as it would arrive at the uterus: a yolk surrounded by a thin layer of very viscous albumen contained and supported within an egg-shaped bag – the shell membrane.

The membrane, which in fact comprises a double layer, is composed mainly of protein mixed with a bit of collagen and is created in the region of the oviduct known as the isthmus immediately before the uterus. Sometimes as you peel a hard-boiled hen's egg you can see fragments of membrane sticking to the inside of the shell. It looks and feels like very thin parchment, but under the microscope you can see that it is a meshwork of fibres. These fibres have been extruded – like Silly String – from thousands of tiny glands in the isthmus region, and laid down to create what under the microscope looks like a piece of coconut matting. This loosely woven structure allows the membrane to expand as the albumen

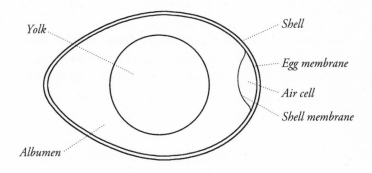

A bird's egg and its main features, outside and in.

is later plumped up with water. The shell membrane is uniform in thickness, and in most birds extremely thin, albeit thicker in larger eggs: about 5μm* in a zebra finch egg; 6μm in a hen's; a robust 100μm in a guillemot's, and 200μm in an ostrich egg. For reference, an ordinary sheet of 80g printer paper is about 90μm thick.[4]

To see what's involved, let's create a bird's eggshell. We will start as the 'egg' arrives at the entrance to the uterus. At this point the 'egg' is like the one I soaked in vinegar – essentially a balloon part-filled with water. In your mind hold it in your cupped hands and imagine that within the skin of your hands are dozens of tiny aerosol sprays of many different types. The first sprays to become active gently extrude a concentrated chalky solution of calcium carbonate that sits like precarious dollops of foam on the balloon surface, each one drying into a lump-shaped meringue. There are lots – probably hundreds – of adjacent aerosols all doing the same thing, so within a few hours the entire surface of the balloon is covered with little dumpy towers of hardened foam (technically they are known as the mammillary cores – so named because of their breast-like shape). Inside the bird, the egg now moves from the eponymous 'red region' of the uterus – a region richly endowed with blood vessels – into the uterus proper, and here another set of aerosols start to squirt water *between* the dollops of hardened foam. The water passes through the surface of the balloon – the fibrous egg membrane – and into the albumen beneath. This process is known as 'plumping', presumably because it plumps up the albumen, and, as it does so, the balloon swells to near capacity. A new set of aerosols are now activated, squirting a concentrated solution of calcium carbonate on top of the foam dollops.[5] This happens all over the egg surface in protracted bursts, so that over about twenty hours they create tall columns packed against each other like a stack of

---

* A micrometre (μm) is a millionth of a metre; a thousandth of a millimetre.

fence posts and referred to as the palisade layer, comprising upright crystals of calcite (a form of crystalline calcium carbonate). As these pillars harden, we have a shell, but not quite. In several places the pillars don't quite meet, leaving tiny vertical spaces. These become the pore canals – the airways that connect the shell membrane with the outside world – so that gases and water vapour can pass in and out through the shell, allowing the embryo to breathe. How the number and size of pores in an eggshell – which differ markedly between species – are determined is completely unknown.

Although about twenty hours have passed since the flaccid balloon entered the oviduct, the process still isn't complete. In the next and final two or three hours, another set of aerosols spray into action, and starts squirting out coloured dyes. The pigment mixes with the last layers of calcium carbonate and creates the ground colour of the egg surface. Once this laying down of ground colour is complete, and sometimes before, another set of paint guns creates the spots and streaks – called maculation – on the egg surface. The way the paint guns produce and project pigments on to the shell is complex and will be discussed in more detail later. The last part of the shell-production process involves yet another group of aerosols that produces a final layer, rather like the coat of wax on a new car. But it isn't wax, it's a sticky protein, and, depending on the species, may also be mixed with some pigment, but it covers the entire surface of the shell and dries almost immediately the egg emerges into the outside world.

Aristotle for some reason believed that the eggshell was soft when it was laid and hardened as it cooled on contact with the air. As William Harvey – who we will meet in the next chapter – explains, this was because a soft-shelled egg would avoid causing the female any pain as it was laid: 'in the same way that they say that an egg softened in vinegar can be easily pushed through the narrow neck of a bottle'. Harvey comments: 'I agreed with this opinion of Aristotle's for a long time until I learned the contrary by infallible experience. For indeed I have ascertained for certain that the egg in the uterus is almost always covered with a hard shell.'[6]

During the twenty-four hours prior to the laying of her first egg of a clutch a female bird is both busy and stressed. Creating an egg requires lots of additional nutrients, but the calcium for the shell is the most difficult to obtain. This is partly because most birds don't carry large reserves of spare calcium in their body, and depend on finding sufficient extra calcium at short notice. The problem is particularly acute for birds like hummingbirds, tanagers and swallows whose normal diet doesn't include much calcium. A colleague of mine who studied barn swallows calculated that because there is so little calcium in their normal diet of flies, if they didn't have an alternative source of calcium, a female would – impossibly – have to forage for as much as thirty-six hours simply to accumulate sufficient calcium to make a single egg.[7]

The amount of calcium needed differs between species and is obviously greater in species that produce a relatively thick eggshell or that produce large clutches, like the blue tit, which can lay sixteen or more eggs. Indeed, blue tits need to find more calcium for their eggshells than is present in their entire skeleton.

Where does this extra calcium carbonate come from?

It comes, ultimately, of course, from what the bird eats. If there's plenty of calcium in a bird's ordinary diet – as there is with lammergeiers that eat little other than bone, and as there is with birds of prey, owls and seabirds like the guillemot, that swallow their animal prey whole – there's no problem. The ingested calcium goes from the gut into the bloodstream, temporarily into the skeleton and then into the glands in the uterus, and from there on to the shell. If there isn't enough calcium in the diet, then a female bird can draw on the calcium in her skeleton, but only a few birds do this. The red knot is one but it has only sufficient stored calcium for two of its four eggs – the rest comes from what it can find during the days it is forming eggshells.[8]

Females in the process of making eggshells search for calcium, and clearly have a specific appetite for it. I find this both remarkable and unremarkable. It is unremarkable in the sense that if they didn't have that appetite, they would not be able to produce adequate eggshells. But it is remarkable that they can tell the difference between calcium-poor and calcium-rich foods, and that this appetite operates only during the period of eggshell formation and usually only in the evening. Given a choice, domestic hens know exactly what they need and greedily go for food supplemented with crushed oyster shell (an excellent source of calcium).[9]

How can female birds tell that they've found a source of calcium? Studies of chickens suggest that there is both an innate and a learned component to detecting calcium. What's not known is what senses birds use to find calcium. Can they smell it? Can they see it? Can they taste it? We simply don't know. Birds bred in captivity, like budgerigars and canaries, are typically provided with cuttlebone, which they can never have seen previously: how do they know to start eating it before laying eggs?

Of a bird's various senses, smell and taste seem the most likely to be useful to find calcium. Humans can detect calcium by smell, but the evidence that we can taste calcium was until recently much less clear. There seems to be some reluctance to admit that mammals (like ourselves) or birds might have specific calcium taste receptors, because this would upset the notion that we possess only a few basic taste receptors (sweet, sour, salt, etc.).[10]

In an ingenious experiment conducted in the 1930s, H. Hellwald fed calcium-deprived chickens on either plain macaroni or macaroni stuffed with crushed eggshell so they couldn't taste the eggshell. Four hours later the birds were given free access to crushed eggshell and Hellwald recorded how much of it they ate. The birds that had unwittingly consumed eggshell hidden in the macaroni ate much less in the second part of the experiment than those that had consumed plain macaroni, suggesting that the birds that had eaten the hidden calcium somehow 'knew' that they had sufficient calcium in their system, without having to taste it. However, this experiment obviously doesn't preclude the possibility that they can taste calcium. Also, as anyone who has been fed by tube will tell you, appetite itself is easily assuaged without experiencing the taste of food. It remains to be shown what senses birds use to detect calcium. In mammals the calcium taste receptor gene has recently been discovered, and the same gene may exist in birds. It is also known that taste buds in different species are sometimes co-opted into doing different jobs, so birds may well be found to have specific calcium receptors in their mouth.[11]

Crossbills are not common where I live on the edge of the Peak District National Park, and my occasional sightings are of birds either flying overhead or sitting in the tops of coniferous trees extracting and eating the seeds from pine cones. As I was looking for information on birds eating calcium I found several mentions of crossbills. The ornithologist Robert Payne reported his surprise at seeing a crossbill in California feeding on the ground, and not just on the ground, but pecking at coyote droppings. In fact the bird – a nest-building female – was picking out the crumbly fragments of rodent bone from the carnivore's faeces. In another account, a group of about fifty crossbills was seen eating mortar. This made me wonder whether the lack of calcium in the crossbill's normal diet of pine seed forces those birds in the process of forming eggs to seek special sources of calcium.[12]

For several species of small sandpiper breeding on the Arctic tundra rodent bones are also their main source of dietary calcium. Egg-laying females acquire the bones and teeth of brown lemmings either from skeletons they find or from the pellets that skuas regurgitate after eating lemmings.[13]

The barn swallows I mentioned earlier obtain their calcium by eating calcareous grit, but most small birds seem to rely on calcium-rich snail shells that they find on the ground during the egg-laying period. Snail searching has been seen in numerous species including great tits, goldcrests and firecrests and North American red-cockaded woodpeckers. The quest for calcium occurs in the evening because eggshell formation takes place mainly at night. Females go to roost with their gizzards crammed with fragments of snail shell whose calcium is extracted overnight and deposited on their eggshell. Experiments with domestic hens found that birds provided with crushed oyster shell in the late afternoon were much less likely to produce defective eggshells than those given oyster shell only in the morning.[14]

As this example shows, and as chicken farmers are all too aware, insufficient calcium can cause havoc with breeding. Defective eggshells are only part of it; without enough calcium birds may lay shell-less eggs enclosed only by the shell membrane, and these, of course, are doomed. With insufficient calcium some birds fail to breed at all. It is easy to imagine careless husbandry resulting in a lack of calcium for poultry or cage birds, but surely wild birds can always find enough calcium?

Not true. In the 1980s Peter Drent and Jan Wijbo Woldendorp in the Netherlands discovered that great tits were struggling to find sufficient calcium to form normal eggshells. The Dutch have several claims to fame, including some of the most intensive agriculture and industrialisation in Europe, which together are responsible for acid rain. This in turn has resulted in the deterioration in soil quality, the loss of woodland and a dramatic reduction in the abundance of snails.[15]

Acid rain was first noticed in the nineteenth century and occurs when pollutants like sulphur dioxide and nitrogen oxide (predominantly from coal-fired power plants) are released into the atmosphere where they dissolve in the water droplets in clouds and fall to earth as rain or snow. The result is the acidification of water bodies, but it also affects soils and vegetation. It wasn't until the 1970s that the full consequences of acid rain became apparent: killing fish, accelerating the deterioration of ancient buildings, and, by removing (leaching, it is called) calcium carbonate from the soil, devastating snail populations.[16]

The lack of snails, particularly in areas of poor, sandy soils where there were few alternative sources of calcium, meant that Dutch great tits and several other small birds produced eggs whose shells were typically 'very thin, granular, porous, fragile and without coloured spots'. In calcium-poor woodlands, female great tits spent a lot of time searching for snails and, unable to find any (or enough), ate grit and sand in desperation. Some females failed to lay at all, others produced defective shells, or occasionally eggs with no shell.[17] The only birds that seemed to be unaffected were those whose territories overlapped with popular picnic sites where they were able to find enough fragments of chicken eggshells from the hardboiled eggs left by untidy picnickers! Intriguingly, the pied flycatchers breeding in the same woods as the great tits had no problem producing normal eggshells. It was initially assumed that this was because the flycatchers were migrants and started egg formation soon after their arrival from Africa, but it was later discovered that pied flycatchers hunt out millipedes and woodlice, whose exoskeletons are rich in calcium. It is not clear why great tits don't do this.[18]

Despite the belated realisation that acid rain was an issue, the negative effects of acid rain on birds' eggshells were found to have been going on – and getting worse – since the Industrial Revolution. This was elegantly demonstrated in a study by Rhys Green and discovered almost by accident. Green, who has a joint position with the University of Cambridge and the Royal Society for the Protection of Birds (RSPB), was looking for possible causes of the decline of British ring ouzels. He wondered whether – as with the Dutch great tits – ongoing acidification of the ring ouzel's upland habitat might have resulted in a reduction in eggshell thickness and breeding success. In contrast, he anticipated that other thrush species like the blackbird, song thrush and mistle thrush breeding in lowland areas – where the effects of acidification were much less obvious – would not be affected in the same way. Using museum collections of eggshells acquired from as far back as 1850, Green was able to track changes in eggshell thickness over time. The eggshells from the ring ouzels showed a continuous decrease in thickness, but so too did eggshells of the other three thrush species. It seems as though they had all experienced the effects of acidification and, critically, the loss of snails.[19]

Changes in the laws relating to industrial and agricultural emissions have reduced the flow of acid rain and both snail populations and the eggshells of tits and thrushes are recovering, but the shells are still thinner than before.[20]

Acid rain was not the only calcium-related problem imposed on birds by our abuse of the environment. A far more serious and insidious issue occurred between the 1940s and 1970s as a result of pesticides. Invented in 1939, the organochloride compound known as DDT (dichlorodiphenyltrichloroethane) proved to be so effective at preventing the spread of insect-borne human diseases that it was said to have helped the Allies win the Second World War. Such was its efficacy that by the 1970s DDT was being employed on a global scale. Its negative effects on wildlife – in the form of dead birds – were obvious almost from the start, but its manufacturers were spectacularly successful in duping or possibly colluding with the US government to use it in sub-lethal doses over vast tracts of land. In the 1940s and 1950s the authorities even pumped DDT on to busy US beaches to keep bathers free from biting flies: a 'service' advertised using the images of carefree children frolicking in clouds of DDT.

DDT accumulates (in a slightly different form, referred to as DDE, dichlorodiphenyldichloroethylene) up the food chain so that top predators like raptors, owls and herons were eventually found to contain high levels. Its effect on these kinds of birds is now well known. Starting in the 1960s birds of prey were found to produce eggs with exceptionally thin shells that the incubating bird invariably crushed. Measurements of eggshells from museum collections showed all too clearly that the reduction in shell thickness coincided exactly with the introduction of DDT. It took rather longer to establish the physiological mechanism responsible, which was that DDE prevented an enzyme important in eggshell formation from working properly. In fact, DDE stopped the secretion of calcium on to the shell sooner than it would otherwise have done, resulting in a thinner than normal eggshell. Unlike the situation in Dutch woodlands, eggshell thinning in the 1960s was not due to a shortage of calcium, but to a chemical that prevented birds from using it. For birds like peregrines, as Mark Cocker in his book *Claxton: Field Notes from a Small Planet* commented: 'The difference between survival and extinction . . . rested on half a millimeter of calcium.'[21]

The use of DDT and other toxic pesticides was eventually banned in the UK and North America in the 1970s and globally in 2001. Raptor eggshells rapidly increased in thickness again, but there is no room for complacency. In 2006 conservationists were delighted to discover a pair of critically endangered California condors breeding at Big Sur on the coast of California. Their excitement was short-lived: no sooner had the eggs been laid than they were crushed in the nest. Their eggshells were very thin and on analysis were found to be loaded with DDE. It was a puzzle. DDT had been banned for forty years – how could this happen? The answer is depressing. Between the 1950s and 1970s the Montrose Chemical Company that made DDT apparently dumped hundreds of tons of it into the Los Angeles sewage system from where it has made its insidious way from the marine sediments into the fish and sea lions on whose carcasses the coastal condors feed.[22]

We have Rachel Carson, and her book *Silent Spring* (1962), to thank for saving us and our wildlife from the greedy, unethical practices of the pesticide manufacturers. Carson, who lost her battle with cancer in 1964, started the environmental movement, but that particular battle is far from over.[23]

In July 2013 I was told about an internet news item claiming that guillemot eggs were self-cleaning. Here was a misnomer if ever I'd heard one. I knew from years of guillemot watching that their eggs were invariably filthy and couldn't possibly be self-cleaning. Intrigued, I looked at the website, and found myself doubly intrigued. It was a report of a paper given at a conference held in Spain by someone called Portugal – Steve Portugal – and was fascinating, first because scientists rarely publicise their work before it has been peer-reviewed and published, and second because of what he had discovered.

Portugal accidentally spilled some water on to a guillemot egg that he had on his desk, and noticed that instead of wetting the surface, the water remained in discrete, silvery droplets. The effect was exactly like that seen on the leaves of the lotus plant – and indeed many other plants. Portugal knew that this was a result of the structure of the egg or plant surface, and for the lotus plant at least is referred to as a self-cleaning effect. The idea is that by forming almost spherical droplets the water holds any dirt away from the leaf surface, and, as the leaf is tilted, the water rolls off carrying away any accompanying dirt.

The microscopic nature of the guillemot's eggshell was something I'd never thought of. As soon as I'd read the article I walked across the corridor from my office to my lab and placed a guillemot egg under my dissecting microscope. At high power, the egg surface looked like the Guilin Mountain range of China: a mass of pointy peaks. I then swapped the guillemot egg for a razorbill's and looked at that under the microscope. For two such closely related species the difference was remarkable: now I was looking at the low rolling hills of England's South Downs. I couldn't believe I'd never thought of looking at eggshells in this way. As a result of his accidental spillage, Portugal had noticed how the pointy mountain landscape of the guillemot egg was very similar to the pimply surface of the lotus leaf, hence the ability of both to force water into tiny spheres. This kind of surface is known technically as hydrophobic (hydro = water, phobic = fearful; that is, repelling).

Portugal's explanation was that guillemots needed a mechanism for coping with salt spray from the sea and for dealing with the fact that there's a lot of detritus around (from other birds) on their breeding ledges. Detritus is what is known technically among seabird researchers as 'shit'. And while it is certainly true that guillemots incubate in filthy conditions, I baulked at the idea that their eggs are self-cleaning. However, the difference between the surfaces of the razorbill and guillemot eggs made me wonder whether the pimply surface of the guillemot's egg is, as Steve Portugal suggests, related to dirt. Razorbill eggs are hardly ever contaminated with faeces because they nest individually and are careful to squirt their liquid excrement away from the breeding area. Guillemots are, by contrast, like carelessly incontinent invalids.

A dissecting microscope isn't the best way of examining an egg's surface. Far more revealing is a scanning electron microscope, which produces lovely, sharp three-dimensional images at higher magnification. I gave some shell fragments to our electron microscope service in the university and a few hours later I had some beautiful images that made the difference between guillemot and razorbill eggshells all the more apparent.

The crisp, black and white images on my computer screen made me wonder about the eggs of the great auk – the extinct, giant relative of guillemots and razorbills. We know that great auks bred in large colonies, but were they packed together in faecal squalor like guillemots or more hygienically spaced like razorbills? Perhaps their eggshell surface could tell me.

But how does one go about examining the surface of a great auk egg? The species is extinct and around the world in various museum collections there are very few eggs. Who would allow me to examine one of these near-priceless specimens? Twenty years earlier I had visited the University Zoology Museum in Cambridge to ask the curator if I could examine one of their eight great auk eggs for another project. He agreed, but as he showed it to me he uttered the same decree my granddad once used when referring to other women: you can look but you cannot touch.

What I really needed was a fragment of great auk eggshell that I could examine using the scanning electron microscope. There are two illustrated catalogues of almost all the known great auk eggs, and it was obvious from the images that over the years one or two have been damaged. This suggested that there might be fragments of shell lying at the bottom of their display boxes that I could have. I wrote to several museums, including the Natural History Museum in Tring, but they all told me the same thing: they had no fragments. Any bits of eggshell were considered untidy and seem to have been thrown out. I was disappointed to say the least, but my other option was to see if a museum would allow me to examine their whole eggs under my dissecting microscope.

I avoided Cambridge, not because they'd refused before, but because they were in the process of moving their museum to a new site and all their specimens were inaccessibly packed away. So I started by asking Douglas Russell at the Natural History Museum at Tring. He agreed – with certain provisos – and a few days later my research assistant Jamie Thompson and I drove from Sheffield to Tring with our dissecting microscope, a camera, a computer and masses of other equipment in the boot of our hire car. Once in the museum we set up a temporary lab on a bench among the museum's seemingly endless egg cabinets. We barricaded ourselves in to avoid any accidental encounters and Douglas gave us a plaster of Paris great auk egg to practise on. Such replicas are not uncommon, and some are exquisitely painted to match particular real eggs; it was a sensible idea to run through our photographic protocol with a dummy egg before looking at the real ones.

Convinced he could trust us, Douglas duly delivered the six precious eggs, each one in its own glass-topped box. They were truly awe-inspiring. So much larger than the eggs of either guillemots or razorbills, their shape was somewhere between the two. Each box was only slightly larger than the eggs themselves, and lying on the cotton wool in which each egg was cradled was a printed label with a brief history of each one. Because these wonderful eggs are so rare, but also so well known, the ownership of each egg has been meticulously – some might say pathologically – monitored.

We looked them over and decided that we would start with the four most perfect eggs, leaving two slightly damaged ones until last.

The first box contains what is known Tristram's egg, originally from Iceland, and probably from the island of Eldey where the last great auk – ever – was killed in June 1844. After several changes of ownership the egg was purchased in 1853 by the ornithologist Canon Henry Baker Tristram who, through his connection with Alfred Newton, was the first to realise that natural selection might be responsible for the appearance of birds and their eggs. This might have made him an evolutionary hero, but after seeing Thomas Henry Huxley trounce Bishop Wilberforce at their now famous religion-versus-evolution debate in the Oxford Museum in 1860, Tristram retracted his belief in natural selection. For many, that meeting was the watershed between god and natural selection as an explanation for the natural world, with Huxley acting as Darwin's mouthpiece and Wilberforce, Bishop of Oxford, batting for the Church. When Tristram died in 1906 his huge collection, including the great auk egg, was purchased by a Mr Crowley who gave it to the Natural History Museum in 1937, where it has been ever since.

We carefully removed the glass top and positioned the box under the dissecting microscope. I was holding my breath. One false move and my ornithological reputation could be shattered. I placed the box under the lens, turned the magnification to its lowest level until the egg surface came into focus, and began to increase the magnification. The image was startling. In an instant I could see that the surface was nothing like a guillemot egg. There were no pointy mountains; instead I saw a patio of flat, crazy paving slabs, much coarser than the razorbill shells we had looked at. In between the patio slabs, I could just make out the openings to the pores. With a warm glow of curiosity and satisfaction we took our photographs and wrote some notes.

Next was Spallanzani's egg. Named for the eighteenth-century Italian priest-cum-scientist, this was – as I mentioned earlier – possibly the oldest egg in the entire museum. It was apparently acquired by Spallanzani in 1760, but from where? No one knew. It eventually ended up in the collection of Lord Rothschild (who in

1901 paid a 'substantial sum' for it) and when he died in 1937 his collection was bequeathed to the Natural History Museum. This is also one of the most beautiful of great auk eggs, with wonderful moss-green pencil squiggles at its blunt end. Viewed through the microscope the surface was similar to Tristram's egg. I was relieved: here was the beginning of some consistency.

Named for its owner, who passed it to the museum in 1949, Lord Lilford's egg is not as attractive as the previous two, being less distinctively marked. Sliding the opened box on to the microscope stage I adjusted the focus, but couldn't quite believe what I saw. The egg surface was completely smooth, marked only by pores that pitted its surface and reminding me of an ostrich eggshell. This was a major blow. It meant that there was huge variation between individual great auks in their egg surface. But at the same time it seemed improbable: I hadn't seen such variation before, but how else could one explain this utterly different landscape? I took a deep breath and, peering again through the microscope eyepieces, started to scan across the egg surface. On and on it went as featureless as the tundra. At one point, though, I noticed a few marks, several of which ran parallel to each other in a short sweeping curve. They were shallow gouges and it came to me in a flash of disappointment and realisation – the egg had been scraped free of its crazy paving. Among the ideas rolling and tumbling in my brain I remembered reading in some of the old books that egg collectors often used corrosive concoctions to remove bird faeces, grime and fungal growth from their eggs.

Just as I was thinking this, Douglas reappeared and asked how we were getting on. When I told him about the scrubbed egg, I could see from his face that he was as disappointed as we were. He disappeared only to come back holding a book on repairing and cleaning eggs. 'Yes,' he said. 'Corrosive sublimate was what the old collectors used: also known as mercuric chloride.' He then explained that because the great auk eggs were so precious and collectors so keen to display their trophies, they made darn sure they were clean and free of fungal growth. What was amazing was that we were the first to notice that some of the Natural History Museum's great auk eggs had been abused in this way.[24] I later found written evidence that eggs were cleaned like this: Symington Grieve in his 1885 monograph on the great auk refers to an egg that was so dirty, its identity remained unknown; it was later purchased by Friedrich Thienemann in the 1840s who, on realising what it was, cleaned it and added it to his collection.[25]

Had these been the eggs of almost any other species, I wouldn't have worried that they had been scraped clean of their cuticle – but great auks! It seemed desperately ironic that in an attempt to celebrate the beauty of their eggs, men had unwittingly discarded an essential, albeit largely invisible part of them.

Of the remaining three eggs, two others had also been scraped smooth – and were useless for our study. The only consolation was that the removal of the paving patio had exposed the underlying pores, allowing us to map their distribution and estimate their numbers. This was some consolation because there is probably no other way we could easily have obtained such information (see below).[26]

A recent account referring to one of the last three great auk eggs we examined says: 'This egg, now faded and broken, can be traced back to the ownership of William Bullock, goldsmith, jeweller and passionate collector.'[27] As was all too clear through its glass top, this was a damaged egg: about one-third of its shell was missing. And before we started to look at the eggshell surface through the dissecting microscope I couldn't help but think that this seemed like a possibility for a fragment for scanning electron microscopy, but I didn't dare suggest it. If the offer came, all well and good; if it didn't then I respected Douglas's curatorial integrity.

We drove back to Sheffield by way of Milton Keynes's dreary road system, chastened by images of corrosively cleaned eggs, but satisfied with what we had been able to achieve with the three good eggs, and immensely grateful for the curator's enthusiastic help.

The next day, after I had downloaded all our photographs and was starting to consider how we would analyse them, the phone rang. It was Douglas. 'Good news,' he said, in a state of obvious excitement. He told me that as he was putting the replica egg back in its correct position in the cabinet, he had noticed a small package in the drawer. It was accompanied by a letter dated 19 January 2001 from Jane Sidell, an archaeologist, thanking the previous curator, Michael Walters, for giving her a piece of great auk eggshell to take some scanning electron microscopy images. Douglas's first response was one of shock. He was aghast and simply could not imagine his predecessor compromising a great auk specimen for this purpose – however important the science. The letter continued, saying that pictures had been taken but remained unpublished. As Douglas was speaking on the phone, I Googled Jane Sidell's name with the words 'great auk' and, although I found her, I could also see that even after thirteen years nothing had been published. Douglas said that the package contained the fragment of shell that I could either borrow to take my own SEM photographs, or he could see if Jane had her original photos to send to me. My next thought squashed my excitement: which egg was the fragment from? I could hear Douglas turning the pages of the letter as he searched for the specimen number. Which one was it? 'Bullock's,' I said (or something similar): and it was indeed Bullock's egg – one of those whose surface had been scrubbed clean. My heart sank, but then I realised it had to have been this particular egg. Taking a piece from Bullock's damaged egg had been *my* first thought on seeing it. Thank goodness I hadn't begged and been granted my wish. Also fortunate was the fact that Jane had, through pressure of work, failed to get round to publishing, for had she done so – and not realised that the egg had been scraped clean – we would have been sent off down the wrong crazy paving path.

Tucked away inside its shell, the embryo has to breathe. But, rather than using lungs, as we do to draw air in and push carbon dioxide and water vapour out, a bird embryo relies throughout most of its development on 'diffusion' – the natural movements of gases, much as do insects (which also don't have lungs). In fact insects and eggs use the same device to do this: tiny pores and pore canals that connect the outside with their interior. For birds there are hundreds

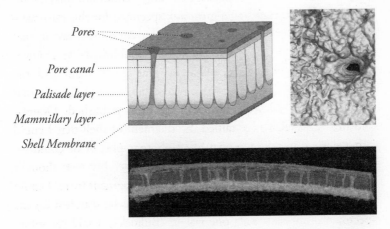

The eggshell. Top left: a 3-D representation of a piece of shell. Top right: the opening of a pore – viewed from above – on the outer surface of a guillemot egg. Lower: a micro-computer tomography image (essentially a micro X-ray) of a piece of guillemot eggshell showing the funnel-shaped pores running from the outside (*top*) to the inside of the shell. The shell is about 500μm thick.

or thousands of tiny pores distributed all over the shell surface. The pores connect, via a narrow tube, the embryo's blood supply to the outside world. Part of the embryo's network of blood vessels lies outside its body and runs under the eggshell, collecting oxygen and releasing carbon dioxide. This structure has an awkward name – it is called the chorio-allantois – and it is analogous to the placenta in mammals.[28]

The pores in birds' eggshells were discovered in 1863 by John Davy, brother of the renowned chemist Sir Humphry Davy. John was a medic and amateur scientist who helped his brother with his experiments. Noted for his 'rather superficial curiosity',[29] John was nevertheless a pioneer in the study of birds' eggs and reported some of his findings at the autumn meeting of the British Association for the Advancement of Science at Newcastle in 1863. Impressed by the enormous variation in the thickness of eggshells in different species, he deduced – entirely logically – that eggshell thickness was related to the weight of the incubating bird. He then noted:

> Whatever the degree of thickness of the shell, it is invariably pervious to air, and chiefly, I believe, through minute apertures – foramina – in the crust [shell] . . . In every instance that I have put an egg under water deprived of air by the air-pump [i.e. under a vacuum] . . . air has been seen to rise in currents from particular points, affording proof of such foramina.[30]

The number of pores per egg varies markedly between species, partly but not entirely related to the size of the egg, with in descending order: about 30,000 on an emu egg; 10,000 on a chicken egg; 2,200 in Cassin's auklet egg and some 300 on the egg of a wren – and, we estimated very roughly, about 16,000 for the great auk's egg. In chicken eggs the density of pores is similar in the middle and at the blunt end of the egg and lowest at the pointed end.[31] Since the pores are fairly straight and run vertically from the inner to the outer surface their length is usually similar to the thickness of the shell. In most species the pores are simple, single tubes, but in ostriches, whose eggshells are very thick, the pores can sometimes have two or three branches. A wren's egg weighing around one gram has pore canals with a diameter of about 3μm; towards the other end of the scale the pores in an emu egg – which weighs 800g – are about 13μm wide.[32]

Generally, the number and size of pores determines how much and how fast oxygen diffuses into the egg. As well as taking away unwanted carbon dioxide, the pores allow water vapour to escape from the developing embryo. As the embryo grows it generates water, referred to as metabolic water and produced as a result of the metabolism of food. It is the same for us, too; we create metabolic water and get rid of some of it at least as water vapour when we breathe. Different types of food generate different amounts of metabolic water; for example, 100g of fat generates – remarkably – 110g of water, 100g of starch creates 55g of water and 100g of protein generates 41g of water.

If the concept of metabolic water seems difficult to grasp, let me tell you about the zebra finch, a tiny bird from Australia that is superbly adapted to surviving in very arid desert conditions, but is more familiar now as a cage bird. In captivity and fed only on standard dry bird seed, zebra finches are able to survive for at least eighteen months without water.[33] They can do this by utilising the metabolic water released as they digest the dry seed. It is this physiological feat that allows zebra finches to persist in some of Australia's driest deserts. It also partly accounts for their appearance in Europe as cage birds in the early 1800s, presumably because they could survive the six-month-long sea crossing to Europe, often, I imagine, without ready access to water.

Inside an egg the developing chick generates plenty of metabolic water from the fat-rich yolk as it grows. This water has to be removed, otherwise the embryo would drown in its own juices, so to speak, and it does this by allowing it to diffuse as water vapour through the pores in the shell. As a result, eggs lose weight during the course of incubation. What is remarkable is that, despite the huge variation across bird species in the size of eggs (in weight from 0.3g to 9kg), in the duration of incubation (10–80 days), and the relative size of the yolk (14 to 67 per cent), the loss of water between laying and hatching is always about 15 per cent of the egg's initial weight. The water vapour lost during incubation ensures that the relative amount of water in the egg is the same in the chick at hatching as it was when the egg was laid. In other words, the composition of the newly laid egg has evolved through natural selection to ensure that the newly hatched chick has the right composition – in terms of the amount of water in its tissues, too. This is achieved by adjusting – via natural selection – the effective pore area such that all the metabolic water produced during development is got rid of before hatching. One consequence of this loss of water vapour is a space in the egg, roughly 15 per cent of its volume, that becomes the air cell at the blunt end of the egg and provides – as we shall see in Chapter 8 – the amount of air needed by the chick just before it hatches.[34]

The air cell is formed between the inner and outer shell membranes when the egg is laid. As the egg cools and its contents contract after leaving the female's body, air is drawn in through the pores and accumulates in a lens-shaped pocket at the blunt end of the egg. If you hold a hen's egg against a bright light you can see the air cell. When you peel a hard-boiled egg the air cell's presence is revealed by the flattened area of white at the blunt end where the air cell has pressed down on the albumen. William Harvey in the 1600s was the first to think about the role of the air cell, dismissing the then widespread belief that its position in the egg signalled the sex of the chick. As development proceeds, the air space increases in size and it is for this reason that you can assess the age of an egg, or its stage of development, from how it floats in water: a very fresh egg with virtually no air cell sinks; older eggs float.

Because gases behave differently under pressure, we might expect the size and number of pores (the effective pore area) to differ among birds breeding at different altitudes. Specifically, the loss of gases will be less at high altitudes. And this is confirmed by a comparison of birds breeding at different elevations: species breeding at high altitudes have fewer, smaller eggshell pores. This could arise because birds adapt to local conditions: in other words birds breeding at different elevations have evolved different effective pore areas, much as animals breeding closer to the poles have smaller ears and other extremities. However, the fact that the same pattern of effective pore area was found in domesticated chickens kept at different altitudes made local adaptation seem unlikely. The acid test, of course, was to examine eggs from the same individuals – in this case chickens again – laying at low and high altitudes. When this was done it revealed that birds are able to detect the difference in altitude and have the physiological flexibility to produce an eggshell whose pores differ in dimensions and number accordingly. This discovery, made by one of the great pioneers of egg and eggshell biology, Hermann Rahn and his colleagues in the 1970s, is one of the most remarkable of all adaptations that birds exhibit.[35] Just consider the possible mechanisms by which this comes about: the bird must be able to detect atmospheric pressure and somehow transmit that, via the brain, to the uterus where the shell is made, to create an eggshell with an appropriate number of pores. Extraordinary!

The pores also allow embryos close to hatching to sense the outside world, in terms of sound and smell at least. Experiments with chickens show that after the embryo has pecked through into the air cell (Chapter 8), but before it has broken through the shell itself, the chick can detect different odours. Embryos exposed to the smell of certain substances at this stage showed a preference for foods associated with that odour when they later hatch.[36] To me at least this seems like a slightly unrealistic experiment with a slightly odd interpretation. It is hard to imagine a brooding parent carrying much odour from the food it has been eating. More probable is the idea that the embryo learns the odour of its brooding parent(s), and later uses this along with several other cues – including calls – to stay close to the adults that will care for it. I can imagine this happening in guillemots, although so far it hasn't been tested.

From the structure of the shell we now move on to consider the rhyme and reason for the shape of birds' eggs.

3

The Shape of Eggs

The shape of the egg usually has a purpose.
O. Heinroth, *Aus dem Leben der Vögel* (1938)

Everywhere I look I can see eggs: blue, green, red and white, but mostly an indifferent khaki colour. Almost all the eggs are intact but some are broken with their orange-yellow yolk and bloody part-grown embryos spilled out on to the rocks. There are eggs piled up in corners, eggs in shit-filled puddles and eggs wedged in crevices. Hundreds, possibly thousands of guillemot eggs have rolled away from where they were laid and now lie cold and abandoned.

I'm on a remote group of seabird islands known as the Gannet Clusters off the coast of Labrador in Canada. I spent three summers here in the 1980s studying guillemots and other seabirds. To my surprise, on this occasion – it is 1992 – I discover that several Arctic foxes have taken up residence on the islands and are causing havoc – killing puffins and frightening guillemots and razorbills away from their colonies. Locals on the mainland tell me that Arctic foxes are common in winter but normally move back north as the sea ice retreats in the spring. This year several foxes have been left behind by the ice and are trapped on the islands for the summer – albeit with a superabundance of food.

The Gannet Clusters comprise six tiny islands, five of which are home to tens of thousands of seabirds: guillemots (of both species), puffins, razorbills, black guillemots, kittiwakes, fulmars and gulls. The common guillemots breed in dense groups on flat rocks close to the sea for there are almost no cliffs on these low-lying islands. We found foxes on two of the six islands, but, curiously, not on the one where the guillemot egg destruction occurred. My guess was that a fox had been there, possibly hopping the few tens of metres between islands on floating ice. I could only imagine the panic as the adult birds scattered in a desperate attempt to escape from the fox that would happily have eaten any of them. Almost miraculously, but also somewhat pathetically, I found two or three isolated guillemots that had returned, found their egg and resumed incubation amid the chaos. With no neighbours to help protect them from predatory gulls and ravens, their chances of rearing a chick successfully were slim indeed.

The sight of so many deserted guillemot eggs left me breathless – I had rarely seen destruction on such a scale. But it also made me wonder about the curiously conical shape of the guillemots' eggs. A more extreme design than that of almost any other bird, the guillemot's egg is said to be *pyriform*, like a pear, in shape.[1] It is barely appropriate because pears come in all shapes and sizes and none has ever reminded me of a guillemot's egg. Sharply pointed at one end, blunt and rounded at the other, whatever we decide to call it – pyriform, conical or pointed – this shape is generally believed to have evolved to stop guillemot eggs from rolling away.[2] My experience in Labrador made me wonder whether there wasn't a better explanation.

Different bird species have a characteristic egg shape that ornithologists variously describe as being oval, spherical, elliptical, biconical or pyriform. These are loose categories, however, and they merge with each other.

When I started to write about birds' eggs I wondered whether anyone had established which of the different shapes was most common. Obviously when we talk about something being egg-shaped we are usually thinking of a hen's egg, which is 'oval', but with an obvious blunt and pointed end and whose greatest width lies closer to the blunt end. To my surprise nobody seems to have quantified egg shape across all families of birds. Part of the difficulty, of course, is coming up with a simple index of shape. Researchers have devised several complicated ways of describing egg designs but there is no single number that captures the full range of shapes. For this reason most books dealing with egg shape simply show – as I do here – a set of outlines or silhouettes illustrating the different types that exist.

One thing we do know is that as well as being characteristic for a particular species, shape is also fairly characteristic for particular families of birds, too. Owls, for example, typically lay spherical eggs;

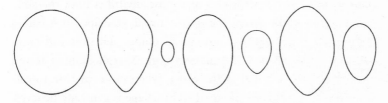

Different shapes of birds' eggs. From left to right: Ross's turaco (spherical), ruff (pyriform), hummingbird (oblong oval or elliptical), crowned sandgrouse (oblong oval or elliptical), African thrush (oval), Slavonian grebe (bi-conical or long subelliptical), alpine swift (ellipitical ovate or long oval). Redrawn from Thomson, 1964.

waders (shorebirds) lay pyriform eggs; sandgrouse produce oval or elliptical eggs; and grebes produce biconical eggs.[3]

As a biologist, two questions come to mind. How are eggs of different shapes made, and why are they the shape they are? The first question is about the mechanics of making an egg; the second concerns the adaptive significance of different egg shapes.

Thinking about how a female bird produces eggs of a particular shape, my natural inclination was to imagine that the shape was determined by the shell: shape and shell created together. The truth is more bizarre. The contours of a bird's egg are governed by its membranes, the parchment-like layer *inside* the shell – as my egg-in-vinegar experiment suggested – rather than by the shell itself. Once you know that the membrane determines the shape it isn't too difficult to imagine the process.

In an ingenious X-ray study of egg formation conducted in the late 1940s, John Bradfield could see that the shape of the hen's egg was determined *before* the shell had even started to form, prior to entering the shell gland. Instead, he could see that the egg's shape was established in the isthmus, the region of the oviduct immediately anterior to the shell gland, where the shell membrane around the egg is created. He noticed, too, that the part of the isthmus adjacent to the shell gland is 'more contractile and more like a sphincter' than the other end adjacent to the magnum, suggesting that: 'Since the egg greatly distends the narrow isthmus [region of the oviduct], it is to be expected that the caudal [tail] end of the egg, situated in the more contractile part of the isthmus, will become more pointed than the cranial [head] end.' He adds, however, that this suggestion is far from proved 'and the problem remains without a clear-cut solution'.[4]

At the end of the egg-shape spectrum opposite to the guillemot are certain owls, tinamous and bustards that lay almost spherical eggs. How is that done? Does the isthmus in these birds lack the sphincter that Bradfield saw in the hen? Or does the egg turn continuously as the membrane is laid down so that the sphincter applies a uniform pressure all over the egg? We don't know.

In humans the maximum size of a baby at birth is determined by the size of the 'birth canal' – that is, the internal diameter of the pelvic girdle. Our present ability to perform caesarian operations removes this constraint, but prior to the twentieth century and the routine use of caesarian section, babies who were too big – or whose heads were too big – failed to be delivered successfully, got stuck and died, usually along with the mother. Strong selection indeed. Because the bones that form the human cranium are still not fused at birth there is some flexibility (literally), permitting the skull to adopt a different shape during birth and allowing some relatively big-headed babies to be born.

Different shapes of eggs may also be how birds produce eggs of a specific volume. If the diameter of a bird's egg is constrained (because the oviduct or the cloaca can stretch only so far), as head size is in human babies, then one way to pack more into an egg is to produce longer, thinner eggs.

The best place to look for evidence of such an effect is in those birds that produce particularly large eggs for their body size: the petrels. The grand old man of petrel biology, John Warham, who died in 2010, provides all the necessary information for making this comparison in his encyclopaedic book on petrels, and it turns out not to be true. The eggs of those species that produce relatively enormous eggs – over 20 per cent of adult body weight – are *more*, not less, rounded than those laying relatively smaller eggs.[5]

It is worth noting at this point that in general smaller birds lay relatively larger eggs, even though in absolute terms their eggs are tiny. The goldcrest weighs five grams and each of its eggs weigh 0.8g (16 per cent of body weight). Even more extreme is the European storm petrel – our smallest seabird – whose egg, at 6.8g, represents 24 per cent of the female's body weight (28g). At the other extreme, eggs that are the largest in absolute terms, like those of the ostrich (weight 100kg) or the even larger (400kg) elephant bird of Madagascar, which became extinct only in the last millennium, produce the smallest eggs relative to body size: in both cases about 2 per cent of their adult body weight. The eggs of each of these four birds are undistinguished in terms of their shape and very similar to that of a hen's egg.

It is fairly clear that neither the size nor the shape of birds' eggs are constrained in the same way that a human baby is. An average human baby weighs about 7.5lb (3.4kg), or 6 per cent of the mother's non-pregnant body weight. If women were to produce offspring weighing 24 per cent of adult body weight like the European storm petrel, their baby would weigh an impossible 30lb (14kg)! The difference is that in birds the pelvis is 'incomplete' and doesn't form a circle of bone as it does in mammals.

This isn't to say that there is no relationship between the design of the pelvis and egg shape in birds. In the 1960s Michael Prynne speculated that egg shape is similar to that of the birds that produce them: long and thin for divers and grebes; rounded for upright-perching owls. Prynne was an oologist whose fifteen minutes of fame occurred on a television quiz show, revealing his ability to make invisible repairs to broken eggshells. He had little scientific knowledge and may or may not have known that twenty years earlier the German zoologist Bernhard Rensch had commented on a similar association, but between egg shape and pelvis shape, rather than bird shape: grebes having a very flat pelvis and elongate eggs, while raptors and owls have a right-angled pelvis. Later, Charles Deeming speculated that while pelvis shape might not dictate or constrain egg shape, it might help to hold a relatively large or conical egg in position within the uterus before it is laid.[6]

What about the adaptive significance of egg shape? Despite several centuries of oological fanaticism we know surprisingly little about why eggs are the shape they are – with the notable exception of some pyriform eggs. For the rest, most ornithologists and oologists have considered shape to be of little evolutionary significance.[7]

Other than guillemots and waders, which we'll come to in a minute, the only other birds that produce really pointed eggs are the king and emperor penguins. Not only do we not know why these penguins' eggs are pyriform in shape, but as far as I can tell nobody even seems to have thought about it. We cannot blame this lack of speculation on their Antarctic inaccessibility, for the birds and their eggs have been known to explorers and researchers for well over a century. I suspect that shape has been eclipsed from the minds of penguin biologists by certain events that took place in 1911 during Robert Falcon Scott's fateful *Terra Nova* expedition in Antarctica. In that year the ornithologist Edward Wilson, accompanied by Henry 'Birdie' Bowers and Apsley Cherry-Garrard, set out from Scott's base at Cape Evans for the emperor penguin breeding colony at Cape Crozier sixty miles (95km) away. Wilson's aim was to obtain developing penguin eggs because it was believed that their embryos held the secret of how birds evolved from reptiles. Emperor penguins breed in the middle of the Antarctic winter and the journey both to and from the colony proved to be extraordinarily demanding. With temperatures as low as minus 60°C, no daylight, blizzards, a lost tent and insufficient food, the three men barely survived. Garrard later described what happened in his book *The Worst Journey in the World*. Three of the five collected eggs survived and the embryos they contained made it back to the British Museum. Tragically after all that effort, the fetal penguins held no scientific secrets and no one asked why the eggshells from which they had been removed were so pointed.[8]

For waders (or shorebirds) such as sandpipers, curlew and whimbrel, the explanation for the conical shape of their eggs emerged in the 1830s. William Hewitson, writing about the common sandpiper, which like other waders invariably lays four eggs, said this: 'as I have mentioned in describing those of the peewit, [its eggs are] admirably adapted by their shape and arrangement in the nest, so as to require the least possible covering: and this is quite necessary, the eggs of this, and all the waders, being remarkably large in proportion to the size of the bird, when compared with those of most other species.'[9]

All waders lay a clutch of four eggs that, when arranged with their pointed ends towards the centre of the nest, fit nicely together, providing the maximum surface for incubation against the bird's

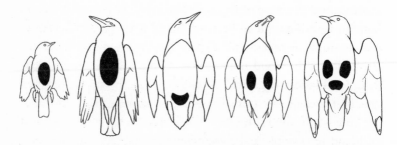

The location and number of brood patches on different bird species. From left to right: Eurasian blackbird (a single large patch), rook (a single patch), common guillemot (a single, centrally placed patch), razorbill (two brood patches even though it lays only one egg; two laterally located patches allows the bird incubate the egg on either side) and a herring gull (three patches to accommodate three eggs).

brood patch. This is a special area of bare skin on the bird's ventral surface that through enhanced blood flow transfers heat to the eggs.

Subsequent tests provide convincing evidence for Hewitson's idea: the pointed shape of wader eggs enhances incubation efficiency because a greater proportion of the eggs' surface is in contact with the parent's brood patch than if the egg was spherical. Pointed eggs also 'provide eight per cent larger eggs for a given brood patch area than would spherical eggs' – in other words, by producing pointed eggs waders are able to create eggs that are 8 per cent larger than if they produced spherical eggs. This is important because the large egg allows for a larger yolk that in turn allows the chicks of wading birds to hatch at an advanced stage of development.[10]

The story of the survival value of guillemot egg shape is a remarkable one. It starts in Scotland in May 1633 when William Harvey accompanied the King of England, Charles I, as he travelled from London to Edinburgh to be crowned King of Scotland. Harvey was the royal physician and rightly famous for figuring out the circulation of the blood and was now trying to unravel the mysteries of fertilisation. Eggs held the clue, and one day in June of that year he travelled east from Edinburgh, taking a boat – without the king – into the Frith (as it was then known) of Forth to visit an island renowned for its abundance of seabird eggs: the Bass Rock.

A volcanic plug rising 350 feet (107m) from the sea, the Bass Rock is an imposing sight. Harvey was thrilled, writing that the island 'shineth with a white glasing and the clifts resemble a mountain of purest chalke'. Crudely dome-shaped, the naturally black rock was covered with the calcium-rich guano of a multitude of seabirds – most notably solan geese (gannets). Harvey likened the rock to an enormous egg encased within its calcium shell.

Although the island was (and still is) most famous for its breeding colony of gleaming white gannets, it was another seabird that captured Harvey's imagination. His guide on the Bass Rock that day told him of a bird even more remarkable than the gannet: a bird that attached its single large egg to a pointed stone. Harvey wrote: 'one bird shewed me above all the rest, which layeth one onely egg, fixing it upon the steep point of a sharp stone'. This was the guillemot, and Harvey must have seen them incubating on the precarious, narrow cliff ledges among the gannets. He must also have been told that the only way a guillemot – which makes no nest – could incubate on such steep ledges was by cementing its egg to the cliff.[11]

The idea isn't quite as preposterous as it now sounds for at least one species of bird – the palm swift – does actually glue its egg on to the palm leaf on which it incubates. Harvey, however, was misinformed, but the story almost certainly had its origins in the fact that abandoned guillemot eggs often become covered in guano and stick to the ledge. The truth about how guillemots successfully retain their egg when breeding on steep and narrow ledges is even more extraordinary.

Harvey's brief mention of the guillemot's egg was followed by other accounts, but discovery and meaning came only slowly. The seventeenth-century Danish priest Lucas Jacobsøn Debes, writing about the massive seabird colony in the Faeroes in 1673, describes how guillemots lay a single egg, 'three fingers breadth' from one another' and that 'when the birds fly away, the eggs rowl often down into the sea'. He makes no mention of their shape however.[12] Another seventeenth-century account, from the pen of Martin Martin – who in 1697 visited St Kilda – seems to be the first to mention the guillemot's unusually shaped egg: 'Its egg in bigness is near to that of a goose egg, sharp at one end, and blunt at the other.' Martin added that the guillemot egg's colour is 'prettily mix'd with green and black; others of them are of a pale colour, with red and brown streaks; but this last is very rare; this egg for ordinary food is by the inhabitants, and others, preferred above all the eggs had here'.[13]

Writing in his *British Zoology* in 1768 Thomas Pennant – one of Gilbert White's main correspondents – thought that guillemots had a trick that allowed them to balance their egg on narrow cliff ledges. He says: 'What is also a matter of great amazement, [is that] they [guillemots] fix their egg on the smooth rock, with so exact a balance, as to secure it from rolling off, yet should it be removed, and then attempted to be replaced by the human hand, it is extremely difficult if not impossible to find its former equilibrium.' Not glued on to a rock, but beautifully balanced.[14]

In 1834 the much-maligned but well-intentioned early conservationist Charles Waterton, squire of Walton Hall near Wakefield, Yorkshire, visited Bempton and arranged with the climmers to be lowered on to the guillemot breeding ledges. He had a good head for heights and twenty years earlier had climbed St Peter's Basilica in Rome (425 feet) and left his gloves on the lightning conductor as a visiting card. The Pope, incensed by Waterton's audacity, ordered him to retrieve them – which he did. In the short essay on his Bempton experience, Waterton notes:

On the bare and level ledges of the rocks, often not more than six inches [15cm] wide lay the eggs of the guillemots: some were placed parallel with the range of the shelf, others nearly so, and others with their blunt and sharp ends indiscriminately pointing to the sea. By no glutinous matter, nor any foreign body whatever, were they afixed to the rock: bare they lay, and unattached, as on the palm of your outstretched hand.[15]

Waterton obviously knew of the egg-stuck-to-rock folklore that Harvey had been peddled, and dispelled it by direct observation. He also knew that there was no balancing trick. Few writers made either mistake again. Surprisingly, though, Waterton's only comment about the shape of guillemot eggs was how variable they were: 'The eggs vary in size and shape and colour beyond all belief. Some are large, others small; some exceedingly sharp at one end, and others nearly rotund.'[16]

In the early 1800s, interest in accumulating collections of eggs was gathering pace and one of the first books in which birds' eggs are accurately illustrated in colour was by William Hewitson. An enthusiastic collector of both butterflies and birds' eggs, he says:

> Were the eggs of the guillemot shaped like those of the major-ity of birds, nothing could save them; their form, which is peculiar to themselves among the eggs of the sea-fowl, is their only protection, it gives them greater steadiness when at rest, and where they have room to roll, the larger end moving around the smaller in a circle, keeps them in their original position: when placed upon the centre of a table and set in motion, they will not wander far.

Read that again: 'the larger end moving around the smaller in a circle, keeps them in their original position'. It is an appealing idea: a gentle knock and the egg spins on its long axis. This is exactly what the TV presenter I mentioned in the Preface did, and I suspect that, like him, Hewitson had ascertained this by using a blown, empty eggshell.[17]

Hewitson's idea was repeated, but without acknowledgement, by the great nineteenth-century populariser of ornithology, the Reverend Francis Orpen Morris. In the 1850s colour printing was just becoming commercially viable and, by working with the printer Benjamin Fawcett and illustrator Alexander Lydon, Morris produced a succession of attractive and very popular natural history books. His knowledge of birds and other aspects of natural history was limited, however, and he was later described as being 'gifted with more energy than was good for him, and a writer of dangerous facility'. In his *History of British Birds* he writes: 'The shape of the [guillemot] egg, which is very tapering, prevents it from rolling off into the sea; for when moved by the wind, or other circumstances, it only rolls round in its own circle, without changing its first immediate situation.'[18] These sound suspiciously like Hewitson's words.

Soon after Morris's book was published in 1856, one of the most perceptive of all nineteenth-century ornithologists, William MacGillivray (1852), had this to say about the guillemot: 'A very little inequality suffices to steady an egg, and it is further prevented from rolling over by its pyriform shape, which however, has not all the effects generally supposed.'[19] Tantalisingly, MacGillivray tells us neither what is 'generally supposed' nor his contrary view . . . but I suspect he's referring to Hewitson's 'spinning on the spot' idea, so facilely plagiarised by Morris.

The notion that the pointed shape of guillemot eggs does little to prevent them rolling away is reinforced by a comment by the Victorian ornithologist Henry Dresser: 'The idea that the birds willfully destroy their own eggs rather than leave them to be taken by the fowler is still prevalent in Shetland.'[20]

Throughout the nineteenth and much of the twentieth centuries guillemot colonies were ruthlessly plundered for their eggs. The enormous colonies in Arctic North America and Russia had probably been raided by local people for centuries, but once these areas opened up to explorers from the south, harvesting of both eggs and adults reached new – and often unsustainable – levels. From the mid-nineteenth century tens of thousands of eggs were taken each year from the colonies in Murmansk and Novaya Zemlya in the Russian Arctic by anyone who happened to be passing. By the beginning of the twentieth century harvesting was out of control with hundreds of thousands of eggs taken mainly for the manufacture of soap, and tens of thousands of adult birds killed for human consumption.

Following the revolution in 1917 and the formation of the Soviet Union, the Bolsheviks banned the commercial exploitation of their Arctic seabird colonies by anybody other than the state. Not only that, the USSR began to study colonial seabirds as part of their totalitarian mastery of the Soviet Union.[21] This extraordinarily prescient move resulted in a succession of biologists – working under what must have been difficult conditions – amassing a huge amount of information on the biology of Arctic seabirds. Their aim was to identify 'the ecological characteristics of each species', with a view to 'conserve and exploit' the birds. In other words, they wanted to know how to maximise their harvest of seabird eggs and adult birds.[22]

One of the most able of these biologists was Lew Belopol'skii who, in his twenties, had been selected to take part in an ambitious Arctic expedition sailing on a commercial steamship, the *Tschluskan*, from Murmansk in the western Soviet Union to Vladivostok on the Pacific.[23] The ship set out in August 1933 but by September was trapped in the ice in the Bering Strait where, on 13 February 1934, it eventually sank. Belopol'skii and all but one of the hundred or so crew managed to escape on to the ice where they established an emergency camp. They also constructed an airstrip – and rebuilt it no fewer than thirteen times – before finally being rescued by the Russian air force in April. On returning to civilisation the expedition leader, the rescue pilots and some of the crew, including

Belopol'skii, received a hero's welcome. They were awarded the Soviet Union's highest decoration that came with a number of privileges which, in Belopol'skii's case, included being made head of the Seven Islands (seabird) Reserve in the Barents Sea. The establishment of this Arctic wildlife reserve coincided with the onset of open-sea fishing, and an important aim of Belopol'skii's work was to use information on the biology of seabirds to enhance the success of commercial fishing.[24]

So far so good. Then, during the Second World War, Belopol'skii was called upon to put his seabird knowledge to practical use. Placed in command of a ship of his own, his main task was to collect seabird eggs and adult birds to help feed the ordinary people of Murmansk, who, because of their remote location and the diversion of what little food there was to the USSR army, faced a desperate food shortage. Belopol'skii's success allowed him to continue to study the biology of Arctic seabirds after the war.

Maximising the number of guillemot eggs that could be taken for human consumption meant minimising the numbers that fell off the ledges. As a result, working out how to reduce the immense number of eggs 'wasted' by falling from the breeding ledges was the single most important issue for Belopol'skii. This in turn was a direct result of what by today's standards were very crude study methods that involved the researchers – rather like the fox in Labrador – walking on to the ledges and flushing the birds, causing their eggs to scatter and roll.

Belopol'skii was well aware, both from his own observations and those of another Russian seabird researcher, Yu Kaftanovski, that the idea that guillemot eggs were like spinning tops was nonsense. Kaftanovski, writing in 1941, had said: 'It is not true that murre [guillemot] eggs resemble tops which merely spin around on the spot at every push or wind movement (as sometimes noted in the popular literature) . . . nevertheless the pyriform shape minimises the fall, especially on uneven terraces where differently shaped eggs would certainly have fallen more.'[25] Belopol'skii writes that, while he agrees with Kaftanovski, he isn't satisfied with his explanation and asks: 'What is the reason for the greater steadiness of pyriform eggs, and under what conditions do they acquire real stability, or reversely, when do they lose their balance and become apt to roll?'[26]

The starting point for Belopol'skii's studies seems to have been the casual observation that most guillemot eggs that rolled off the ledges and smashed on the rocks below contained tiny embryos, suggesting to him that freshly laid eggs were more vulnerable to rolling off the ledges than well-incubated eggs. To check this Belopol'skii and his colleagues conducted an experiment in which they gently pushed eggs that had been incubated for only a few days: they all rolled off the cliff. They later conducted the same experiment with eggs that had been incubated 'for a longer period' and found that 'in the latter case the eggs, as rule [sic] literally roll along the curved circumference [i.e. in an arc] and remain on the ledge'.[27]

By today's standards their descriptions of these experiments are extremely vague and hardly convincing – not least because with so little detail it is impossible to verify exactly what they did and how many eggs they tested.

Belopol'skii describes how he then asked a colleague, Savva Mikhaïlovich Uspenski, to conduct what he calls a 'mass experiment' in which guillemots that had only recently laid their eggs were frightened from their breeding ledges by a shot being fired. In response to the gunshot, Belopol'skii says, the eggs started 'pouring' – which is hardly surprising. The researchers then went back to the same ledge when the eggs were close to hatching and repeated the experiment: bang! 'An enormous swarm of birds rose in the air, but not a single egg tumbled over the ledge'.[28]

Again, because they present so few details it is hard to judge the value of this 'experiment'. However, Belopol'skii fails to discuss the most obvious reason for the fact that in the second part of the experiment no eggs 'tumbled over the ledge' – it seems pretty obvious that all the vulnerable eggs had already been lost when they scared the birds off their newly laid eggs.

Intriguingly, in seeking an explanation for this apparent effect Belopol'skii noted that, as incubation proceeded, the centre of gravity of a guillemot's egg altered such that the blunt end of the egg was raised up from the rock surface on which it was laid, compared to eggs incubated for less time. This effect is a consequence of the air cell in the blunt end of the egg increasing in size during the course of incubation (see page 46 in Chapter 2). The shift in the centre of gravity results in a well-incubated egg rolling in a tighter arc than a fresh egg – and hence being (somewhat) less likely to roll off a ledge.

Belopol'skii implies that this shift in the centre of gravity is an adaptation that enhances the stability of the egg. However, since the eggs of all bird species undergo a similar change, regardless of their shape, it is unlikely to be an adaptation, and more likely a consequence of the change in the centre of gravity. However, it is possible that the effect is more pronounced in a pointed egg, and hence does enhance its stability.

The Russian biologists' view of what constituted an adaptation in guillemots was coloured by their study methods, which included walking on to the breeding ledges, firing shots in the colony and generally causing panic among the incubating guillemots. Undisturbed, guillemots almost never leave their eggs unattended, and rarely in a panic. Only in the presence of humans or predators such as polar bears and Arctic foxes – both relatively rare events because of where the birds choose to locate their colonies – does such disturbance occur, often with considerable loss of eggs through rolling. I find it extraordinary that Belopol'skii and his colleagues continued to think in Darwinian terms at all given that the teaching of evolution in Stalinist Russia was dominated by the erroneous Lamarckian philosophy so aggressively promulgated by the Soviet director of agricultural science, Trofim Lysenko.[29]

Belopol'skii's years of seabird research culminated in the production of a substantial volume entitled *Ecology of Sea Colony Birds of the Barents Sea*, published in 1957, with an English edition in 1961. When I was a research student in the 1970s Belopol'skii's book was an important source of information. However, its poor production, tiny figures, muddy photographs, slightly shaky translation and nasty, thin paper all conspired to devalue the science for me. Having reread it recently, however, my opinion changed. I recognised that science was done differently then, and that Belopol'skii's achievement was utterly remarkable and far in advance of seabird studies conducted around the same time elsewhere. Quite apart from the biology, among the most striking things I noticed on rereading his book were his negative comments about both Lysenko and the Soviet political system.

It was only when I delved into Belopol'skii's past that I discovered the reason for this. In 1949, as part of the so-called Leningrad affair in which a paranoid Josef Stalin suspected high-ranking communist colleagues of treason, Belopol'skii's brother, wife and father were arrested. Jealous and suspicious of the young communists in Leningrad, Stalin fabricated cases against them. Belopol'skii's brother, who was head of a holiday resort for the communist elite, was (falsely) accused of spying for Britain. Stalin had him shot. Three years later, in 1952, Belopol'skii himself came under suspicion – presumably guilty by association – and was thrown out of the Communist Party. Even though Belopol'skii's status as a national hero of the *Tschluskan* expedition was supposed to give him political immunity, the system had a way of dealing with that: summoned to court, he was forced to sign away this immunity. Thus stripped of his privileges, Belopol'skii was sent to a gulag in the Novosibirsk region of Siberia (east of Omsk and some 500km north-east of Astana, the capital of present-day Kazakhstan) for five years, in what the authorities considered a light sentence. The charge? Guilty of being his brother's brother. Fortunately, after Stalin's death in 1953, Belopol'skii was released and rehabilitated. In 1956 he established the Rybachy Biological Station – a bird observatory – on the Curonian Spit, on the Baltic, a major bird migration route. He died in 1990 aged eighty-two or eighty-three.[30]

The Russian scrutiny of guillemot egg shape was inconclusive and the 1950s saw a new era of guillemot research. It started in 1956 when a Swiss biology teacher, Beat Tschanz (pronounced 'chance'), together with another biology teacher and a curator from the natural history museum in Bern, Switzerland, made a three-week visit to the seabird island of Vedøy, in the Lofoten archipelago off the Norwegian coast. Tschanz was beguiled by the guillemots and, although already in his mid-thirties, decided to return to university and study for a PhD on their behaviour. Sadly, I never met Tschanz, but I was intrigued by the fact that someone from a country with no coastline and no seabirds should choose to study guillemots.[31]

Tschanz was fascinated by the fact that guillemots breed at such high densities, with no nest and on precarious cliff ledges. The focus of his research was to determine how guillemots cope with such an unusual breeding environment, or, more precisely, what adaptations they possess that allow them to breed in this way.

In the late 1950s, the study of adaptation was starting to become topical and a similar study on cliff-nesting adaptations in the kittiwake gull was already underway by Esther Cullen, one of Niko Tinbergen's students at Oxford.[32] Tinbergen, who was later awarded the Nobel Prize (together with Konrad Lorenz and

Karl von Frisch) for developing the study of animal behaviour, had conducted research over several years on the visual and vocal displays of gulls. The kittiwake was one of many gull species he and his students studied, but it was the only one to breed on cliffs. By comparing the behaviour of the kittiwake with that of other gulls, all of which are ground-nesting, Cullen was able to identify the adaptations that enabled kittiwakes to nest successfully on tiny cliff ledges.

Tschanz's guillemot studies, which continued over several decades, focused on three main topics, two related to the problem associated with high-density breeding: (i) recognising your own egg (Chapter 5), and (ii) recognising your own chick (Chapter 8); and a third, associated with breeding on narrow cliff ledges with no nest: egg shape. During the mid-1960s Tschanz made several trips to England to meet Tinbergen, who was impressed by Tschanz's egg-and-chick-recognition studies – and at Tinbergen's suggestion even conducted experiments on black-headed gulls while they were together. Like Tinbergen, Tschanz used the comparative approach, comparing guillemots with some of their closest relatives, mainly the razorbill, but also the puffin and black guillemot to identify adaptations. Even though the Russians, as we have seen, had already done a lot of research on guillemot egg shape, Tschanz was keen to conduct his own independent studies and to re-evaluate and extend some of their ideas.[32]

The first results from Tschanz's research on guillemot egg shape, made in collaboration with Paul Ingold and Hansjürg Lengacher and published in 1969, provided clear evidence for the adaptive significance of the pointed egg shape. They categorised guillemot eggs according to shape – which, as Waterton had pointed out in 1834, is remarkably variable – as slightly pointed, moderately pointed and extremely pointed. They then looked at whether eggs of each shape would roll off a cliff ledge when pushed. For comparison they also included the much less pointed eggs of the closely related razorbill. The result of these experiments was that the more pointed the egg, the less likely it was to roll off the cliff ledge.[34]

In writing about these results a Dutch ornithologist, Rudi Drent, later observed:

> The applicability of these tests to the natural situation was verified by allowing parent *Uria* [guillemots] to incubate dummy eggs of plaster on the ledges, and observing the decrement [losses] with time. *Uria* eggs survived better than *Alca* [razorbill] types, in the most extensive observation series significantly so (42/50 [84 per cent for the guillemot eggs], compared to 35/50 [70 per cent for the razorbill eggs]).

These results seemed to provide conclusive evidence that the pointed shape of the guillemot's egg is indeed an adaptation that makes it less likely to roll off the breeding ledge.

But Drent, who was an extremely perceptive biologist, seems almost to have been predisposed to accept Tschanz's conclusions. The difference is not actually statistically significant – despite what Drent said – and most biologists would have been rather more cautious in accepting that the difference was biologically convincing.[35]

They would have been right to be cautious. In fact, Tschanz and his colleague Paul Ingold subsequently recognised for themselves that those initial experiments, despite their seemingly clear-cut results, were not as convincing as they seemed. First, eggs made of plaster behave differently from real eggs (they weigh less and the weight distribution within the egg is very different from that of real eggs); second, they realised that the substrate itself had an important effect on egg rolling; and third, they recognised that the parent birds played a significant role in retaining the egg on the ledge.

Ingold started again, and, to cut the story of a long – forty-seven page – paper short, this is what he did and discovered. He compared the rolling behaviour of real guillemot and real razorbill eggs on ledges with a variety of natural substrates and found that guillemot eggs were *no less* likely to roll off than razorbill eggs. In rolling experiments on an artificial (relatively smooth) surface with different inclinations, guillemot eggs – because of their more pointed shape – rolled in a smaller arc than razorbill eggs. However, on a natural, much more uneven, surface there was no difference in the rolling arc of the two species. This was a consequence of the razorbill's egg being lighter in weight than the guillemot's egg. Thus – and this is the critical and subtle point – because guillemot eggs are larger and heavier (110g) than razorbill eggs (90g), guillemot eggs *would* be at greater risk of rolling off, if they were the same shape as razorbill eggs. In other words, given that the guillemot has to produce an egg of a certain size, the pointed shape does provide some protection from rolling.

In addition, and inevitably, Ingold showed that the steepness of the slope and the nature of the surface – smooth or pebbly – has an important effect on an egg's propensity to roll. He also noted striking differences in the incubation behaviour of the two species, with guillemots incubating more assiduously, with fewer and shorter breaks than razorbills, which may or may not be an adaptation in guillemots for keeping the egg safe and less likely to be lost when partners exchange incubation duties.

The fact that the weight of the egg influences its rolling behaviour suggested to Ingold something that had always puzzled me – that Brünnich's guillemots, despite always breeding on narrower ledges and hence having eggs with an even greater risk of falling, lay *less* pointed eggs than those of (common) guillemots. Indeed, many Brünnich's guillemot eggs are more like those of razorbills in shape.

Ingold suggested that because Brünnich's guillemot is, on average, a smaller bird than the common guillemot, and therefore lays a smaller egg (weighing about 100g), it can get away with producing a less pointed egg.

If Ingold was right, there was a way we could test the rolling-in-an-arc hypothesis: we could compare the shape, specifically the degree of pointedness between different populations of both guillemot species. If the rolling-in-an-arc hypothesis was correct the prediction was that the larger and heavier the egg, the more pointed it would be. Like many bird species guillemots tend to be larger the further north they breed, and larger birds tend to lay larger (and heavier) eggs. The latitudinal increase in body size is known as Bergmann's rule, after the nineteenth-century German anatomist and physician Carl Bergmann who suggested that larger animals had a smaller surface area to volume ratio and hence stayed warmer at cooler latitudes.[36]

Museums have large collections of eggs of both guillemot species, and from different parts of their geographic range – that is, from different latitudes – so it didn't seem too difficult to collect the data necessary to test this idea. Over several months, my research assistant and I visited most of the main European museums, photographing and measuring over one thousand guillemot eggs. After all that, we found barely a shred of evidence for Ingold's idea. First, although the Brünnich's guillemot eggs were on average significantly less pointed than the common guillemot eggs (as was already known), their eggs had exactly the same volume (effectively the same weight as when fresh) as those of the common guillemot, which meant that Ingold's idea fell at the first hurdle. Second, larger eggs (in both species) tended to be more pointed – as Ingold predicted – but to such a minuscule extent that it is probably biologically irrelevant.[37] These results strongly suggest that pointedness probably serves some purpose other than rolling-in-an-arc. The mystery of guillemot egg shape remains a tantalising biological puzzle.

Lupton's enchantment with guillemot eggs was driven by the extra-ordinary: mainly in terms of colour and pattern, but also in shape and size. Among his collections are drawers of dwarf and giant guillemot eggs. Dwarf or runt eggs have been known for as long as people have kept chickens. They are rare and in the past their appearance was often associated with various superstitions, includ-ing the notion that were laid by cockerels – and for this reason they were sometimes referred to as 'cock eggs'. They were sometimes also called 'wind eggs' because of an ancient fantasy that they were ferti-lised by the wind blowing up the hen's oviduct – which was wrong on both counts and not least because such eggs were invariably unfertilised. Runt eggs typically have no yolk and arise because the infundibulum fails to catch the ovum as it is released from the ovary (see page 24). Without the bulk of the yolk, the oviduct produces a miniature and yolkless egg. On other occasions a tiny piece of tissue sloughed off the oviduct wall fools the oviduct into starting the egg-making process, but, with no yolk, a dwarf egg ensues.[38]

Less often, hens produce very large eggs, which on being opened are found to contain two yolks. Double-yolkers are relatively enor-mous precisely because they contain two yolks.[39] They usually occur when two ova develop simultaneously in the ovary and are released at the same time. They are rare: in forty years of fieldwork I only once found – in Labrador – a double-yolked guillemot egg. At Bempton, however, double-yolked guillemot eggs *seem* to have been relatively common. Rickaby mentions them several times in his diary, and recounts how two double-yolkers were collected from one section of cliff on the same day![40] Lupton loved double-yolkers and over the years acquired a staggering total of forty-four, including one he was especially proud of that weighed six ounces – that is 170g, substantially more than the average Bempton guillemot egg of 110g. A 170g guillemot egg would be the equivalent of a woman producing a 12lb baby – difficult, but not impossible.

While two embryos often develop within doubled-yolked eggs, there are very few records of twin birds of any species hatching and surviving from such eggs – there simply isn't enough albumen (see Chapter 6).

In chickens the incidence of double-yolked eggs is around 1 in 1,000.[41] Assuming that some 10,000 guillemot eggs were collected each year at Bempton, and if the incidence of double-yolked eggs is the same (and we don't know this) in guillemots and chickens, then the climmers might have expected to come across about ten a year.

As well as extra-large and extra-small examples, Lupton's collection includes several malformed eggs whose shapes fall outside those occurring naturally in almost any bird species. These guillemot monstrosities range from almost spherical runt eggs, through elongate, tube-like pointed eggs, elongate symmetrical eggs, to asymmetrical mango-shaped eggs. The only other bird where we know of a similar range of bizarre egg shapes is the chicken and that's not surprising given that worldwide there are six billion laying hens producing a trillion eggs each year.

Lupton's unusually shaped guillemot eggs demonstrate that the bird's oviduct is capable – under certain circumstances – of producing an egg of almost any rounded shape. What we do not know is whether any of these oddly shaped guillemot eggs would have hatched had they been left. I imagine that rather few would have survived the full period of incubation. I also suspect that their presence in Lupton's cabinets was because of the climmers' regular visits to the ledges. Indeed, I think that the climmers themselves may have been partly responsible for the extraordinary array of misshapen, missized and miscoloured guillemot eggs that were obtained from Bempton. Perhaps the incessant disturbance disrupted the birds' egg formation process. In all my years of studying guillemots at relatively undisturbed colonies I have seen a runt egg only twice, and have never seen a seriously dumpy or grossly asymmetric egg.

Of all the misshapen guillemot eggs in Lupton's trays the ones that intrigued me most were those that resemble a mango: slightly flattened on each side and with a distinct curve to their form. If you were trying to design an egg not to roll and not to fall off a ledge, this is it. The fact that a female guillemot *can* produce such an asymmetric egg suggests that if such eggs resulted in a chick there may have been selection for it.

Let's finish by looking at ordinary 'egg-shaped' eggs. Can it really be true, as some ornithologists have suggested, that for the majority of species egg shape has little or no selective advantage?

In the past it was thought that the shape of the egg was somehow dictated by the form of the chick that would emerge from it. This is what Fabricius said in the 1600s: 'The eggs of almost all birds, as a matter of fact, are not perfectly round, but elongate . . . because the chick is longer than it is broad; again the egg is not exactly oval and uniformly elongate, but more obtuse, broader and thicker at one end . . . since the chick is broader at its upper end where the head and thorax are situated.'[42] He goes on to discuss the variation in egg shape in chickens, picking up an old idea on the relationship between egg shape and the sex of the chick inside it, and that relatively broad eggs gave rise to females, because he believed (incorrectly) that chickens, like women, are broader at the hips than cockerels and men. William Harvey despaired of his tutor: 'The reasons which Fabricius advanced for the shape of the eggs, I am glad to pass over for they are all invalid.'[43]

Obviously, for virtually all organisms the main constraint on egg shape is that it must be circular – more or less – in cross section so that it can be formed and transported along the oviduct. But do eggs have to be symmetrical along their long axis?

The fact that for birds 'egg-shaped' implies *not spherical* is important. The eggs of most other animals, such as fish, frogs and marine turtles, *are* spherical. This suggests that the slightly elongate shape of birds' eggs, and those of reptiles like snakes, lizards and crocodiles, has some selective advantage. There are several possibilities.

The first relates to the fact that a sphere has the smallest surface-to-volume ratio of any shape. Any deviation from spherical therefore means that the surface area is greater in relation to the egg's volume, which in birds may be important for increasing the efficiency of heat transfer from the brood patch to the egg. Of course, a non-spherical egg will also cool down more rapidly when not being incubated. The ovate form of most birds' eggs may be a compromise between how effectively it can be warmed when it's being incubated and how rapidly it cools when it's not. In this respect, shape may also be influenced by the average number of eggs in a clutch and by the shape and number of brood patches the species has.[44]

Second, since the eggs of many reptiles – which do not incubate their eggs through contact with a brood patch – are also elongate, another factor may be important. The most likely one relates to 'packaging'. We don't know for certain whether snakes, lizards and crocodiles are constrained in the diameter of their eggs, but they may be for they do have fairly elongate bodies. However, for large crocodilians that produce relatively small but elongate eggs, it seems unlikely that the body sets a maximum diameter for egg size.

A third constraint for birds' eggs must be the strength of the shell: eggs must be strong enough to take the weight of the incubating adult yet weak enough for the fully developed chick to break out at hatching. Again, a spherical egg must be optimal in terms of supporting the weight of the incubating bird, but perhaps a slightly elongate egg, by providing the chick with more leg room and therefore more leverage, aids the chick's emergence from the shell when it is time to hatch.[45]

It is remarkable that after so much egg research in past decades, there are so many questions for which we don't yet have an answer.

Let's see whether our understanding of the colour of birds' eggs is any better.

4

Colouring Eggs – How?

. . . But where a passion yet unborn perhaps
Lay hidden as the music of the moon
Sleeps in the plain eggs of the nightingale.
                    Alfred, Lord Tennyson, *Aylmer's Field* (1793)

On one occasion when I was visiting the Natural History Museum in Tring to look at their collections of eggs, I fantasised about being there with George Lupton, the two us standing side by side in front of an open drawer containing dozens of guillemot eggs. I ask Lupton what he sees. *Sheer beauty*, he says: *the form, the size and, above all else, the diverse yet harmonious palette of colours and patterns.* He then asks me what I see. *Data*, I reply, or, rather, *data lost*, for most of the eggs in front of us have no cards and no information. I then add that I'm not indifferent to the aesthetics of eggs, but as a scientist I think first of what those eggs could have told us about the lives of birds had they had data cards. But I think also of what museum eggs can still tell us: we have not yet finished interrogating dataless eggs. Lupton then turns and asks me what I mean by *data*. It is a rhetorical question for he is familiar with several scientifically articulate collectors, but Lupton's question gives me pause for thought. What *do* I mean by data?

Data are the bits of information I use to interpret the natural world. That's what scientists do: that's their goal. I want to understand why an owl's eggs are white; why guillemot eggs are so variable in colour; why thrushes' eggs are blue, and why the eggs of some tinamous are a startling grass-green. We scientists gain our understanding by making observations – looking at stuff, just as Lupton and I are looking at this tray of eggs – and asking, *why this? Why that?* Lupton tells me he's done that himself, so I think that makes him a scientist, too. The difference is that, for me, asking the question isn't enough on its own; I need to go one or several steps further. *Why this?* becomes *perhaps this*. That is, I've formulated an hypothesis, a statement of what I imagine might be the explanation.

Tinamous lay the most extraordinary and beautiful eggs – like glazed porcelain – and those of different species are blue, green, pink or purple. If I'm serious about answering the question, I then think of ways of testing my hypothesis by subjecting it to the most ruthless scrutiny. I ask myself, what would destroy my hypothesis? What I most certainly do not do is search specifically for evidence that would support it. If my hypothesis is still okay after subjecting it to the most rigorous test, I can start thinking that we understand why tinamou eggs are the colour they are.

To come up with a sensible hypothesis you have to know something about the birds' biology. George Lupton knows a tinamou egg when he sees one, but he hasn't been to Central America and he's never seen a nest, so he's not in the best position to think up a great hypothesis. I have seen tinamous in the wild, and I even supervised a student who studied tinamous, and I know that their eggs are laid on the ground on damp leaf litter and when not covered by the incubating bird (in tinamous, it is usually the male) they glow brilliantly in the forest half-light. Why so conspicuous? One hypothesis – and, actually, it's the only one I can offer Lupton – is that tinamou eggs are distasteful and unpalatable so their luminescence is a warning: 'Don't eat me, I'll make you sick.' It is a relatively easy idea to test, too. Unconvinced, Lupton nods thoughtfully and mutters, *Hmmm, perhaps*. Pausing for a moment, he adds: *Perhaps tinamou eggs are brightly coloured and shiny because they are made from a different form of calcium carbonate than the eggs of other birds*.[1]

Now it is my turn to pause. Lupton's explanation or hypothesis is fundamentally different from mine. It isn't an alternative. The difference is subtle, but crucial to understanding the way scientists interpret the world. Our two hypotheses are equally valid, but they are like looking at the world through slightly different lenses. My hypothesis is about the evolutionary significance – the adaptive significance – of egg colour, and asks how the bright, shiny shell of a tinamou egg increases its chances of it becoming an adult tinamou. It is the same as asking *why* tinamous produce eggshells like this. Lupton's question, on the other hand, is a *how* question: *how* do tinamous produce that immaculate gloss? His question is concerned with the mechanics, or processes, of making an eggshell. Both approaches are equally valid, but they are distinct and need to be thought of separately, at least initially. Muddling them up simply causes confusion. My hypothesis is as incapable of answering the question *how* as Lupton's is of explaining *why* tinamous produce such eggs, and, while the mechanistic question can richly inform the evolutionary one, the reverse isn't always true.

It is for this reason that biologists are so keen to distinguish between these two approaches to understanding the natural world. Of course, it wasn't really until Darwin drew our attention in the mid-1800s to natural selection as the mechanism of evolution that we could even contemplate *why* questions, although a few perceptive individuals had previously started to think in this way, admittedly with God's wisdom as the final explanation.[2] It is also worth saying that while both types of question are equally valid, and most biologists recognise that they need to know about *how* and *why* to fully understand a system, the reality is that science is so big that to make progress you have to specialise and that usually means focusing on either *how* or *why*. The two types of question have also differed in their popularity over time. Following the revolution in our understanding of natural selection with its focus on 'selfish gene thinking' in the late 1960s and early 1970s, *why* questions have become much more attractive to researchers than *how* questions.[3] This in turn means that there has recently been more research funding for *why* questions. There is also a sense that the *how* questions have all been answered, so it is more productive, more exciting and more rewarding to address the unanswered *why* questions.

I am a firm believer in needing to know the answers to both types of question, and in this chapter and the next I'm going to tell you about the *how* and *why* of egg colour, respectively. Here we will focus on *how* questions, then, in the following chapter, we will look at our current ideas that might explain *why* eggs are the colour they are.

I was surprised – and rather heartened – to discover that one of the first people to investigate the chemical nature of egg pigments was also instrumental in founding the University of Sheffield, where I've spent most of my career. Like many other mid-nineteenth-century scientists, Henry Clifton Sorby was independently wealthy. He was also a polymath, perhaps best known for recognising that the addition of carbon to steel created a much stronger product, an idea that helped to make Sheffield the 'steel city'. Marine biology was one of Sorby's other interests and he developed an ingenious method of rendering organisms such as worms, jellyfish and ctenophores as two-dimensional lantern slides without disrupting their shape or structure.

Sorby was also fascinated by colour and in the 1870s devised a microscope and method for identifying the pigments found in eggshells. In fact, eggshells were just one of dozens of coloured biological materials he analysed in this way. The principle was similar to the flame test, which I did at school, a method based on the fact that different substances when burned emit a different coloured flame, allowing the substance to be identified. Sorby realised that as well as emitting different colours when burned, different substances also absorb light differentially and that this could also be used to identify them.

Spectrum analysis, as it was called, was a hot scientific topic in the 1860s, its efficacy demonstrated at the time by its use in the discovery of two new chemical elements. What Sorby realised was that spectroscopy could be adapted for use with a microscope: 'By interposing a solution or transparent substance between the light source and an arrangement of prisms, the black lines caused in particular parts of the observed spectrum by the absorption of light rays by the substance could be used for purposes of identification.'[4]

By taking pieces of eggshell and dissolving away the calcium, Sorby was able to leave the colour in solution. Projecting a light through the solution and looking at the colours that didn't get through – represented by the black bands – he could deduce their composition.

Prior to Sorby, several researchers had thought that the colour on eggshells was an accidental by-product of the egg-laying process, for example due to the leakage of blood through the wall of the uterus, or to contamination by bile pigments. Some even thought that the colour and markings on eggs were due to the unintended contamination by faeces as the egg passed through the cloaca.[5] This latter idea was first aired in the early 1800s and over the next century had a number of enthusiastic supporters. It is remarkable how persistent the idea was even after cuckoo enthusiast Eduard Opel in the 1850s and guillemot enthusiast Henry Dresser in the 1870s reported finding perfectly coloured eggs in the oviducts of the birds they dissected . . . that is, before these eggs even entered the cloaca.[6]

Sorby's careful analyses demonstrated that eggshell colour was determined by substances that were also in blood and bile, but did not produce those colours in the way that many of his predecessors thought. He identified seven different pigments, and concluded: 'So far as I am able to judge from what is now known, the colouring of eggs is due to definite physiological products, and not to accidental contamination with substances whose function is altogether different.'[7] Essentially what he had previously deduced about the colours in plants turned out to be true of eggs, too, and was due to the 'relative and total amount of a limited number of definite and well marked substances'.[8]

Sorby gave his seven eggshell pigments some horrible names, although they are logical and decodable, most obviously those with the prefix 'oo' which simply means eggs: (1) oorhodeine, (2) oocyan, (3) banded oocyan (the band here refers to a band seen in the spectrum during analysis), (4) yellow ooxanthine, (5) Rufous ooxanthine, (6) an unknown red and (7) lichnoxanthine.

We don't need to worry about these names because, even though subsequent studies gave them different and equally difficult names, there are in fact really just two classes of eggshell pigments, protoporphyrin and biliverdin, both of which are involved in the synthesis and breakdown of haem, the iron-based substance that gives red blood cells their colour. I realise these names are tricky, too, but there are only two of them.

Porphyrin, which is responsible for reddish-brown colours, is generally referred to as porphyrin-IX (the 'IX' refers to an aspect of its chemical structure) to distinguish it from the other porphyrins that are so widespread in biological materials they have been referred to as 'pigments for life'.[9]

Biliverdin was what Sorby called 'oocyan' and is responsible for the blue-green colour of eggs (oo = egg, cyan = blue/green). It is a breakdown product of haemoglobin, and is also responsible for the green colour sometimes visible in bruises.[10] It is worth pointing out that it is the mixing together of these two pigments that gives eggs their wonderful tonal quality, which, of course, contributes to their enormous visual appeal as natural works of art.

I was intrigued by how long Sorby's ideas took to penetrate the zoological literature, and how J. Arthur Thomson, writing in his *Biology of Birds* in the 1920s – forty-five years after Sorby's discoveries – continued to believe that the colours were waste products 'excreted' on to eggs, and therefore non-adaptive:

It is very likely that they [colours] are unimportant by-products or waste products of the bird's metabolism, which are got rid of along with the all-important nutritive secretions from the wall of the oviduct.[11]

Thomson tries to strengthen his case by drawing a botanical analogy:

The pigments in withering leaves are very beautiful and very striking, but, so far as we know, they are devoid of biological significance except as end-products and by-products in the essential chemical routine of the green leaf.[12]

Commenting on the variability in egg colour, he says:

If it should become of survival-value that the egg of a guillemot or a cuckoo let us say, should settle down to some particular coloration, there is plenty of raw material on which the process of natural selection could work.[13]

Summing up, Thomson writes:

> Our argument is that there is no need to search too diligently
> for the utilitarian significance of the distinctive coloration of
> birds' eggs. The pigmentation may be a by-play of metabo-
> lism and the constancy of pattern an expression of an orderly
> constitution. And that may be all.[14]

The study of eggshell pigments took a step forward in the 1970s,
when Gilbert Kennedy and Gwynne Vevers worked out the relative
amounts of the two main pigments in the eggshells from 106 bird
species. White eggs, with no obvious colour visible to us, proved
to be particularly revealing: those of the northern fulmar, Eurasian
dipper and ring-necked parakeet contained neither pigment; those
of the white stork, scops owl and European roller contained protoporphyrin but not biliverdin, while the eggs of certain penguins and
the woodpigeon contained both.[15]

We now know that there can be pigment in any or all of the layers
that make up the shell. In certain species even the shell membrane
is pigmented; in others the calcium layer of the shell is coloured,
and in others like the domestic fowl pigment is confined to the
cuticle – the very thin outermost layer. In some birds, including
raptors, some pigment spots are embedded deep in the shell and are
invisible from the outside. One researcher found that by dissolving away the shell surface these pigment spots 'gradually come into
view, disintegrate and float away'.[16]

In the rock and willow ptarmigan, not too distant relatives of
the domestic fowl, of course, the entire egg surface is wet on being
laid and my colleague Bob Montgomerie, who studied ptarmigan
in Canada, told me that it wasn't unusual to find feather marks on

eggs where the laying bird had brushed the egg soon after laying. In addition, if he picked up a newly laid egg, his fingerprints were left on its surface. On top of that, both the ground colour and the maculation oxidise and change from a reddish to a browner tint within twenty-four to forty-eight hours of laying.

Intriguingly, Wilhelm von Nathusius states in a study from 1868 that the pigment on newly laid guillemot eggs is, like that on ptarmigan eggs, wet and smudgeable. I wonder how he knew. I've never come across any other mention of this and one might have imagined that the Bempton climmers would have commented on it. Nor have I noticed it myself.[17]

In the past researchers also occasionally found a white egg in the uterus of a species that normally laid coloured eggs. This is what happened to Friedrich Kutter in 1878 with a common kestrel he dissected. He also noted tiny specks of red-brown pigment on the surface of the uterus. Now, you might imagine that he would have put two and two together and deduced from this that the egg was just about to have its colour added and that the uterus was the source of colour, but he didn't. Instead, he presumed that the pigment specks had travelled down from the ovary. Some subsequent researchers followed his lead, but, as we've seen, there was in the nineteenth century a variety of views on the way eggs acquired their colour. As late as the 1940s, Alex and Anastasia Romanoff in their egg-biology bible *The Avian Egg* wrote, 'All evidence points to the uterus as the site of pigment secretion', and in a review written in the 1970s Allan Gilbert, another egg expert, said that although it was now clear that the pigment glands must be in the uterus their identity was still unknown.[18]

In fact, in the 1960s Japanese researchers studying Japanese quail had, using a microscope, examined the cells — they are referred to as epithelial cells ('epi' meaning 'on' and 'thelial' meaning nipple,

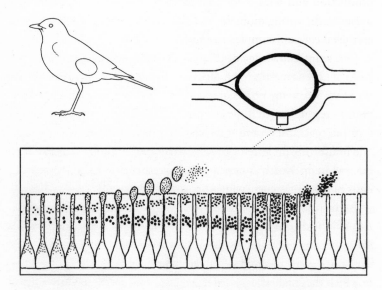

A scheme showing the way pigment may be applied to the eggshell in a bird's uterus. Top: an outline of a bird showing the location of the egg within its body; right: then after the egg in the uterus; lower: the pigment-producing cells lining the uterus are basal cells (with a broad base) and apical cells (with a narrow base). The production and release of pigments into the oviduct and onto the shell reads from left to right. Redrawn from Tamura and Fuji, 1966.

because the term was first used to refer to tissue with a nipple-like surface) – lining the uterus and seen tiny globules of pigment lying within them waiting to be released, like paint from a tiny paint gun, at the right moment. This research also showed that pigment production and release are very precisely timed. If you examine a uterus at the wrong moment – before or after pigment release – the epithelial cells are completely empty.[19]

Our scant knowledge of the egg-colouring process is entirely the fault of the domestic fowl. Because chickens lay unmarked eggs, there was never any financial incentive – or much opportunity – for poultry researchers to study the production and deposition of markings on eggs. They were interested in the ground colour, because 'the housewife' preferred eggs of a particular colour – brown in the UK, white in North America – and so there was research into the basis of that difference, which proved to be genetic.[20]

In the past as I walked to my guillemot study colonies on Skomer I usually had to pass through several gull colonies – there were so many herring gulls and lesser black-backed gulls in the 1970s they were impossible to avoid – and it wasn't uncommon to see clutches in which the smallest, last laid egg was a pale blue colour rather than the typical mottled khaki of the other two eggs.

Something similar occurs in the Eurasian sparrowhawk where the last one or two eggs of a clutch (which can comprise as many as seven eggs) are often much paler and less heavily marked than those laid earlier in the sequence. In addition, the replacement clutches of sparrowhawks, laid after the first is lost, are always relatively pale.[21]

The interpretation is that the female has run out of pigment. If this really is the explanation, it raises the question of how important egg colour is anyway. Another possibility is that the paler late and replacement eggs constitute an adaptation of some sort. Of several suggestions, one is that last laid eggs are less valuable and serve as a decoy egg for potential predators; or that early laid eggs are more vulnerable because they are not incubated by the parent for several days, and therefore need better camouflage. Neither of those explanations seems very convincing. Another idea is that odd-coloured last eggs signal to potential brood parasites, like cuckoos, that the clutch is complete, and therefore not worth parasitising. That doesn't seem very plausible either because a brood parasite would then benefit by eating or destroying the eggs to make the host lay a replacement clutch.[22]

Yet another idea is based on the fact that light penetrating the eggshell has a positive effect on the developing embryo. Somewhat surprisingly, the eggs of domestic fowl and Japanese quail hatch sooner if the embryo can detect a certain amount of light through the shell.[23] Whether this is true in other birds has yet to be established, but it does provide a possible explanation for later laid eggs being less heavily marked. Such eggs may allow more light to reach the embryo and hence speed up its development – which may be an advantage with later hatched gull chicks within a clutch, or replacement sparrowhawk clutches. However, to be convinced we would need to check: (i) whether less heavily marked eggs do hatch sooner, (ii) that the more rapid development is a direct result of the embryo receiving more light, and (iii) whether the time this saves is really long enough to make a difference to survival. This could also be another factor favouring white eggs among cavity nesters. Slightly less obviously, it may also provide an explanation for the long-standing mystery of why so many open cup nesters – like thrushes – lay light blue eggs. It is blue light that has the greatest accelerating effect on embryo growth in poultry. Since the colour of the eggshell dictates the wavelength of light reaching the embryo, a blue eggshell may maximise the amount of blue light reaching the embryo and minimise the duration of incubation, and hence the eggs' vulnerability to predation.[24]

One of the really astonishing things about guillemot eggs is that within a single species there is almost every possible colour and type of marking that one can find among all of the ten thousand other species of birds. Wallace's comment that guillemots have gone crazy with their eggs is true. One of George Lupton's egg-collecting contemporaries, George Rickaby, left a diary of his oological exploits in which, as we saw in Chapter 1, he included illustrations of twelve main types of guillemot eggs, described by their names, including *pepper pot*, *shorthand*, *scrawl* and *nose cap*. These marks can occur on almost any colour background, although analysis of the colour and type of marking on Brünnich's guillemot eggs revealed that this association is not random, and eggs whose ground colour was dark tended to have larger markings than eggs with light ground colours.[25] Rickaby's classification of egg types must have developed among the climmers and collectors for easy reference. If a mere twelve types covered the entire span of variation in guillemot eggs, that would have been easy, but in fact as he points out the variation is virtually endless, as indicated by the additional terms the collectors used: *green petal*, *thick brown undermark*, *super red*, *blue pepper and salt*, *pencil in shell*, and so on.[26]

Although millions of pounds and dollars have been spent on research on eggshell formation and structure in the domestic fowl, until recently almost nothing was known about how marks were made on the shells of other species. Studies of Japanese quail, which produce heavily marked eggs, reveal that the markings are applied in the final three or four hours before laying. In those species like raptors and guillemots, where some of the markings lie deeper in the shell, colour must be produced earlier than this. In many guillemot eggs it is often obvious that patches of pigment have been overlaid with more calcium and more ground colour – as Nathusius noted in his studies from the 1800s.[27]

In Chapter 2 I said that the uterus was full of glands operating like miniature paint guns to apply pigment to the shell. Put like this it sounds straightforward, but the more I think about it, the less satisfying this explanation appears. The colour, pattern and distribution of marks on guillemot eggs are often extraordinarily complicated. The best way to appreciate just how complicated is to try making a convincing coloured illustration of a guillemot egg. I have attempted this and it is tricky. I have also had to paint model guillemot eggs for some of my experiments and matching a model egg with a real egg is far more difficult than I imagined. I have immense admiration for those artists (usually uncredited) who created convincing images of the eggs of guillemots and other species in some of the early oology books. Painting your own guillemot model egg is also an instructive exercise when trying to imagine just how those patterns are applied to a real egg.

For some guillemot eggs – such as Rickaby's *black cap*, with nothing more than large black blobs covering the larger end of the shell – there's no problem. I envisage the egg lying immobile within the shell gland with a few broad-nozzled paint guns directed at the blunt end of the egg and firing away until the covering is complete. Similarly for a slightly more complicated type of egg, known as *pepper pot* – comprising a uniform scattering of tiny coloured marks evenly distributed over the shell – it seems relatively simple. Here, I imagine thousands of minute paint guns evenly distributed within the shell gland, firing for just a few moments to create this simple but lovely speckled pattern.

Much more difficult to imagine are the heavily patterned eggs, whose marks bleed or blend into one another. I'm thinking here of a design more commonly seen in razorbill eggs, but which very

occasionally occurs – as Lupton's collection shows – in guillemot eggs. In these eggs the colours seem to have been laid down
in layers, each one successively overlaid with calcium, to create a
graded appearance. Not only that, the colours on these eggs seem to
be smeared in a wonderfully controlled way. Perhaps as the various
paint guns are firing the egg is rotated slightly within the shell gland,
creating an effect much as you might by smudging wet paint – very
carefully – with your finger.

The most bizarre and intriguing, and possibly the most revealing, patterns of all are the types of marks Rickaby calls *shorthand*
or *scrawl*, which I have always called *pencilling*. These are eggs
with white, cream or light blue ground colour decorated with
what appear to be the random, seemingly endless calligraphy
from a brown or black paint gun. In some cases, it seems as
though only a single paint gun is firing; in others there appear
to be several.

I have struggled to imagine how such marks might be applied.
The most likely possibility is that once the paint guns are firing,
the egg rotates within the shell gland in such a way that these
scrawls appear to be randomly applied. Imagine artist Jackson
Pollock trickling paint across a canvas from a heavily loaded
brush, or a drunk carrying two or three leaky cans of paint staggering around a smoothly concreted courtyard. However, if this
is the way pencilling is achieved we might expect to see some
concordance between marks made by the different paint guns. In
many such eggs there is no obvious concordance, although this
has not been subject to mathematical scrutiny. I'd love to collabo-
rate with someone who could check this. To do so, I imagine
that we might 'peel' the egg, lay its pattern out flat and map the
different lines, and check for similarities that might then reveal
the secret of application.

Not being a mathematician, nor – so far – finding one to do this, I tried something else. Using hens' eggs, I assumed for simplicity that the shell gland had just three paint guns, and positioned three differently coloured felt tip pens in separate laboratory clamps, pointing inwards with a gap of about three centimetres between them. Taking the egg by its pointed end, I held it among the pens so that their tips were in contact with the eggshell and then rotated the egg. As I hoped, pencil-like markings appeared on the shell but, contrary to my expectations, it was hard to detect any 'concordance' between the three colours.

The other thing that is obvious about real pencil markings is their uniform width (about a millimetre) and how clean they are. I said earlier that the colours on razorbill eggs often appear to be smudged, yet the pencil markings on guillemot eggs are anything but smudged; there's a sharp edge, and no seeping into the ground colour. Now, if you were using a paint gun yourself to create similar markings on a surface, you would need very quick drying paint. You would also need the equivalent of a felt-tip marker rather than a can of spray paint, or a big brush full of watery paint. If the guillemot's egg really was rotating inside the shell gland, without very rapidly drying paint it is hard to see how the lines avoid becoming smeared or smudged.

There are a few other birds that produce pencilling patterns on their eggs. Certain bowerbirds that breed in Australia and New Guinea, including the spotted, greater and in particu-lar the yellow-breasted (aka Lauterbach's) bowerbird, produce wonderfully marked eggs, as do jacanas or lilytrotters, moor-hen-like birds of the tropics. Closer to home, the European yellowhammer, once known as the 'scribblyjack', lays eggs in which the dark pencil lines vary in thickness and, in contrast to guillemot eggs, often seep evocatively into the ground colour, reminiscent of the animal images – executed in a mixture of blood and earth – that decorate caves at Font-de-Gaume, Niaux and Rouffignac in France.

The poet John Clare caught it perfectly:

Five eggs, pen-scribbled o'er with ink their shells
Resembling writing scrawls which fancy reads
As nature's poesy and pastoral spells –
They are the yellowhammer's and she dwells . . .[28]

Perhaps the eggs of guillemots and yellowhammers move a great deal in the shell gland as the colour is applied. To produce some of the observed patterns these movements would have to be quite complex, and probably in all directions. Another possibility is that, rather than the egg rotating within the shell gland, it remains still, and the paint guns move. Are there free-ranging paint devices that crawl over the shell depositing colour as they go? Or perhaps there are areas of shell – linear or otherwise – predisposed to pick up pigment? Both ideas seem improbable, but at this stage we need to keep an open mind.

Sometimes we can learn a lot by looking at systems when they go wrong. Lupton's passion for guillemot eggs included being enamoured of several colour types that we now know – from studies of chickens – were errors. One of these is a type he and other oologists called *banded*: eggs with a pale or unpigmented band around their middle. In chickens, and presumably other birds, too, this is the result of some kind of shock while the colour is being applied to the shell. The fact that this manifests itself as a band two or three centimetres wide in a guillemot egg suggests that the colour and pattern are applied sequentially along the length of the egg rather than simultaneously all over the egg.

Another pattern, referred to as *nose cap* by Rickaby, is an egg whose pointed tip is darkly pigmented rather than the more usual blunt end. Nose caps might occur if the distribution of the dark paint guns are reversed in these birds, or they may be eggs that failed to turn over just before laying – as the eggs of other birds are known to do (see pages 179–84). If the typical blunt end concentration of colour occurs because the dark, broad-barrelled paint guns lie at one end of the shell gland, then it is easy to imagine how an egg that fails to turn could end up with colour at the 'wrong' end.[29]

For now the creation of complex patterns on the eggs of several bird species remains a future project.

5

Colouring Eggs – Why?

With a strange perversity, many who have discussed this question seem to find insuperable difficulties in explaining the lack of protective colouration which obtains in the great majority of coloured eggs. But in such cases, almost without exception, these eggs are deposited in a nest, itself not infrequently a conspicuous object.

W. P. Pycraft, *A History of Birds* (1910)

Our exploration of the evolution of egg colour begins, not with Darwin as you might expect, but with Alfred Russel Wallace. The co-discoverer of natural selection, Wallace has always played second fiddle to Charles Darwin. In 2013 on the centenary of his death huge efforts were made through scientific meetings and television programmes to resurrect Wallace's scientific reputation. I'm not sure how successful this was but at the very least his name is better known now than it was.

The discovery of natural selection by Wallace and Darwin provided a completely new way of explaining the natural world. First recognising the phenomenon in the 1830s after returning from his momentous *Beagle* voyage, Darwin spent the next twenty years thinking through its consequences before daring to publish. Wallace, however, came to natural selection late, in 1858.

It is now well known that Wallace, then in Indonesia, wrote to Darwin to tell him about his idea. Deeply shaken, Darwin was extremely anxious about priority – his priority – and wrote to his close colleagues Charles Lyell and Joseph Hooker to seek advice. They knew that Darwin had been beavering away at natural selection for a long time but decided that under the circumstances the fairest thing to do was a joint public presentation of both men's ideas. They arranged for Wallace's essay together with a synopsis of Darwin's idea to be read before the Linnean Society in London on 1 July 1858.

> The accompanying papers, which we have the honour of communicating to the Linnean Society, and which all relate to the same subject, viz. the Laws which affect the Production of Varieties, Races, and Species, contain the results of the investigations of two indefatigable naturalists, Mr. Charles Darwin and Mr. Alfred Wallace.[1]

It was a joint performance at which neither of the principal players was present and the meeting passed almost without comment. Few of those at the meeting seemed to recognise the significance of what they'd just heard. To Darwin it was a relief since it gave him some breathing – and writing – space, prompting him to stop dithering and put pen to paper to produce *The Origin*, which was published the following year.

Wallace seemed happy to remain in Darwin's shadow and spent much of the rest of his life exploring the extent and significance of natural selection. He investigated a wide range of topics and described his various findings and ideas in no fewer than ten books. Perhaps the most intriguing of these, published in 1889, seven years after Darwin's death, has a simple, deferential title: *Darwinism*.

Wallace and Darwin had a shared vision about the efficacy of natural selection, but they disagreed over Darwin's singular idea of sexual selection. Puzzled by the fact that males and females often differed in appearance and behaviour, Darwin came up with the ingenious idea of sexual selection. In a nutshell, sexual selection can favour brightly coloured plumage in male birds, for example, if the survival costs are offset by increased reproductive success. Put another way, although the long and gaudy tail of the male Argus pheasant, for example, may compromise its ability to escape from predators, if it makes him irresistible to females, he will leave more descendants than a less gaudily plumaged male. Evolutionary success is measured in descendants.

Darwin envisaged sexual selection occurring through two processes: competition between males for access to females, and female choice.[2] He also assumed that the starting point for sexual selection was the dull plumage similar to that seen in the females of many birds such as the mallard duck, and that males exhibiting a bit of colour were rewarded by greater reproductive success. In other words Darwin imagined sexual selection favouring the evolution of brightly coloured males.

Wallace disagreed. First, he didn't attribute the same importance to sexual selection as Darwin did, and, second, he assumed that bright plumage was the starting point and that selection favoured the evolution of dull plumage in females. The reason for Wallace and Darwin's different viewpoints was that Wallace thought that the bright colour of males was due to their greater vigour. Indeed, Wallace assumed that the bright colours in males were a mere by-product of this greater vitality and that selection favoured dull, cryptically coloured females because when incubating their eggs they were better able to avoid detection by predators.

Of sexual selection Wallace said: 'The only way in which we can account for the observed facts is by supposing that colour and ornament are strictly correlated with health, vigour, and general fitness to survive.'[3]

Wallace and Darwin's polite debate over sexual selection and the evolution of bright colours continued for years. Although the subsequent consensus was that Darwin was right, Wallace made some important discoveries.[4]

One of Wallace's findings concerned the colour of caterpillars. This had been a conundrum for Darwin who recognised that their often bright hues could not possibly have evolved through sexual selection because caterpillars cannot reproduce (it is only the adult forms, the butterflies and moths, that do that). Wallace correctly deduced that bright colours in certain caterpillars have evolved to protect them from predatory birds, their bright colours advertising the caterpillar's unpleasant taste.[5]

A second puzzle that Wallace addressed was the colour of birds' eggs. This was a topic Darwin had not considered – even though his grandfather Erasmus Darwin had commented on bird egg colour in his book *Zoonomia*.[6] The puzzle was this: how could the often conspicuous colours of eggs be adaptive? Wallace thought hard about this and started to see some patterns – so to speak – including the possibility that egg colour might, as Erasmus Darwin suggested, confer some protection from predators. First, Wallace said, eggs can be divided into two groups: those that are white and essentially uncoloured, and those that are coloured. Because eggshells are made of calcium carbonate – carbonate of lime, as he called it – which is white, the 'primitive' colour of birds' eggs must have been white, just as it was in their reptilian ancestors.[7] He then went on to point out that all those birds such as kingfishers, bee-eaters, woodpeckers, trogons and owls that lay their eggs in concealed places produce white eggs. Wallace proposed that under such circumstances – that is, when eggs are invisible to the outside world and hence immune to the forces of natural selection – white eggs are of no disadvantage and have therefore retained their ancestral, uncoloured state. Although the link between cavity nesting and white eggs is often attributed to Wallace, I wonder whether he got the idea from Hewitson, who as far as I can tell was the first to mention it.[8]

The implication of Wallace's explanation for white eggs is that where eggs are visible their colouration is likely to be adaptive. Before discussing that, Wallace had to deal with those birds that breed in the open yet produce white eggs, and therefore appeared to contradict his idea. Certain pigeons, nightjars and the short-eared owl breed in the open but, he says, their white eggs are rarely left unattended, kept safe from predators because the parent birds themselves are protectively coloured. Although that seems sensible, it might still be adaptive if their eggs were cryptically coloured for those few occasions when they are left unattended – unless, of course, white colouration is advantageous. We shall return to this later.

Continuing, Wallace says: 'We now come to the large class of coloured or richly spotted eggs, and here we have a more difficult task, though many of them decidedly exhibit protective tints or markings.' He then identifies a number of birds, including the little tern and ringed plover, whose eggs so perfectly match the shingle on which they are laid that they provide excellent protection from predators like crows and foxes.

Getting to guillemots, he says that their eggs provide an example of exactly the opposite, in terms of their brilliant conspicuousness and – as Wallace thought – in being invulnerable to predators. This is what he writes:

The wonderful range of colour and markings in the eggs of the guillemot may be imputed to the inaccessible rocks on which it breeds, giving it complete protection from enemies . . . [The eggs] have become intensified and blotched and spotted in

the most marvellous variety of patterns, owing to there being
no selective agency to prevent individual variation having full
sway.[9]

In other words, Wallace believed guillemot eggs to have been under
no selection pressure and as a result to have been able to evolve the
remarkable variety of colour we now see. 'Having full sway', as he
calls it, is part of Wallace's view that egg colour, like plumage, is
determined by 'vigor' and that different species generate colour as
an inconsequential by-product of their vitality . . . and that unless
checked, as it is in the ringed plovers and terns by predation, it
runs free. Wallace's thoughts here on how natural selection works
are reminiscent of a child whose behaviour is held in check by
its parents, but once outside their control runs off skipping and
playing in wild exuberance. As we'll see, for all his genius Wallace
was wrong. He was also wrong about the white eggs of cavity
nesting birds being free of natural selection. He was subsequently
shown to be mistaken about guillemot eggs being indifferent to
natural selection because of what he considered their predator-free
breeding sites. For Wallace, a perceived lack of natural selection
produces no colour in one case and a riot of colour in the other,
which sounds to me as though Wallace wanted to have his oological cake and eat it. At the very least he could have argued that
guillemots are generally more vigorous than cavity nesting birds,
but he doesn't.

One of the reasons natural selection has been such a challenging
concept for many non-scientists, and for some scientists, too, is that
it can often be difficult to see how it works. For example, J. Arthur
Thomson, whom we met in the previous chapter, revisited Wallace's
white egg ideas in the 1920s, writing:

Perhaps a deeper way of looking at it [the issue of white
eggs] is to say that the absence of shell-pigment is a primitive

feature, as seen in reptiles, and that birds which persisted in having white eggs had to seek out concealed places or build well-covered nests.[10]

This is typical of the kind of evolutionary argument used in the 1920s before natural selection and genetics came together in the 'modern synthesis' to explain biological phenomena. By today's standards Thomson's argument seems rather muddled: it presupposes that egg colour is 'fixed', but the choice of nest site isn't. In other words, that the choice of nest site is more amenable to natural selection than egg colour. We wouldn't make such a presumption today.

We now know that natural selection operates on variation, for example in egg colour, and that this variation must be coded within the genes, without which there can be no evolutionary change. We also know that variation arises as a result of genetic mutations, and if Thomson was posing his white-egg argument today, he might say that: (i) in some species there may never have been the mutations causing differences in egg colour that natural selection could act upon, and (ii) in such species selection must — instead — have favoured other traits, such the choice of more concealed nest sites, the cryptic plumage of adult birds or stronger nest defence behaviour by parents.

There are several different hypotheses for the association between white eggs and concealed nesting cavities. One, suggested by Alexander Morison McAldowie in the 1880s, is that the pigment in eggshells protects the developing embryo from solar radiation – and hence is unnecessary among birds nesting in dark cavities.[11]

As Wallace recognised, the eggs of the first birds were probably white and unmarked like those of their reptilian ancestors. Reptiles do not need coloured eggs because they conceal their eggs in the ground, in cavities or under vegetation, and so they are relatively safe from both predators and solar radiation. As birds evolved and began to use different types of nest site, it was inevitable that their eggs would be exposed to daylight and therefore more vulnerable to predators. It seems reasonable to suggest, as both Erasmus Darwin and Wallace did, that under such circumstances brown and speckled eggs would be less conspicuous and less vulnerable to predators.

There are two ways we could test Wallace's hypothesis that coloured and patterned eggs are safer from predators than white, immaculate eggs. First we could conduct a comparative study, and compare the colour of eggs produced by birds breeding in different habitats and see whether any broad patterns emerge.[12] Second, we could perform an experiment, and the most obvious one is to alter the colour of eggs and see whether that affects their chances of being taken by a predator. There have been many such tests, often using chicken eggs painted either white or cryptically coloured and left in artificial nests whose vulnerability to predation is then monitored. This seems such an obvious experiment, whose outcome is so apparent it hardly seems worth doing. Yet the results overall – and there have been quite a few such experiments – provide little support for Wallace's idea. In most cases, white eggs are no more susceptible to predation than cryptically coloured eggs. Does this mean that Wallace was wrong? Maybe. But it is also possible that the experiments were wrong, for several reasons. First, painted experimental eggs may have smelled as well as looked different, and thereby attracted mammalian predators who hunt mainly by scent. Second, because birds' vision is different from our own, it is possible that, despite the researchers' best artistic endeavours, their cryptically painted eggs were not very cryptic to avian predators like crows and magpies. Third, many of these experiments involved placing painted eggs in artificial nests which themselves may have been conspicuous.

For those who like the idea that the colour of birds' eggs provides protection from predators, blue eggs, like those of the European

song thrush or American robin, which are startlingly conspicuous
in their brown nest cups – at least to the human eye – have always
been a problem. Erasmus Darwin assumed such eggs to be cryptic
because seen from below through what he called their 'wicker nests'
they were less visible against the blue sky. This is wrong at several
levels, including the fact that nests of species such as the song
thrush and American robin are lined with mud and impossible to
see through; and that he assumed that the main predators would see
the nests from below.

Overall, there are three broad categories of evolutionary explanation for the colour of eggs, and we will deal with each in turn:
camouflage and conspicuousness, avoiding brood parasites and
individual recognition.

If you've ever walked on a shingle beach amidst a screeching horde
of sandwich terns, you'll know how difficult it is to avoid stepping
on their wonderfully camouflaged eggs. You may even have been
unlucky enough to experience the dreadful sensation of an egg
giving way under your foot. The eggs of many ground-nesting birds
like terns, plovers, waders and quail are often so perfectly camouflaged that such a mistake is all too easy. It is not difficult to imagine
the way that natural selection over successive generations perfected
the similarity between egg and background. Eggs that don't quite
match their background are discovered and eaten, and genes responsible for that mismatch simply don't make it to the next generation.
Exactly the same phenomenon occurs in the peppered moth *Biston
betularia* – the classic example of natural selection in action. On
the soot-stained trees created by the Industrial Revolution, selection
favoured moths that were darker, and hence better camouflaged
from predatory birds. Subsequent research on peppered and other

moths showed an important additional aspect to their camouflage: as well as natural selection continuing to refine their camouflage over successive generations, natural selection also favoured moths that chose to settle on backgrounds that maximise the effectiveness of their camouflage.[13] This raises the question of whether birds like terns or waders choose to nest in those places that provide the best camouflage. A recent study of Japanese quail, whose eggs – as we've seen – are often heavily marked, found that females seem to 'know' their own egg colour and select their nest site to provide the best possible match. This behavioural component greatly increases the effectiveness of the camouflage. The big question is how female quail *know* what kind of egg they lay – do they learn this at their first breeding attempt or is this knowledge innate such that egg colouration and choice of breeding habitat are linked?[14]

What about the spots and streaks on the eggs of those birds that breed in dark cavities – what are they for? One group of researchers thought they had found the answer when they discovered that pigment spots (composed mainly of porphyrin) seemed to occur most often on thin points on the eggshell, suggesting that they might be there to add strength, compensating for low calcium levels in certain parts of the shell. Subsequent research by others, however, found no evidence for this idea. We are still in the dark when it comes to explaining the significance of spotting among birds that breed in the dark.[15]

An explanation for the bright colours of certain eggs is that they evolved specifically to be conspicuous. This idea was first proposed in the early 1900s by Charles Swynnerton, an avid egg collector who disagreed with much of what Wallace said about egg colour. Swynnerton suggested that brightly coloured eggs might be distasteful and might therefore have evolved, much like noxious insects, to be seen and hence avoided by predators. He tested his idea by presenting eggs of a number of different birds to several (tame) mammalian predators including a rat, a bushbaby and an Indian mongoose. He also used humans as subjects and recorded their response to the taste of eggs. One of his correspondents, Mr H. M. Wallis, described how, instead of blowing eggs (to obtain the shells) he sometimes sucked them and found that the flavours varied immensely: 'Thus, robin, nightingale and swallow are beastly. But the white eggs of the little bittern are sweet and mild as cream.' Swynnerton's tame mongoose also seems to have considered only certain eggs to be tasty, avidly eating the blue eggs of the dunnock and blackbird but rejecting the white eggs of the wren and great tit.

Swynnerton was a careful observer, and commendably keen to *test* (rather than prove) his hypothesis that conspicuously coloured eggs were unpalatable. He was also smart enough to recognise the limitations of his numerous experiments, and in the end decided that there was little evidence for any link between the colour and palatability of eggs.[16]

Despite Swynnerton's conclusions, the zoologist Hugh Cott revisited the idea that brightly coloured eggs were distasteful thirty years later, in the 1940s. Cott was consumed by the association between conspicuousness and palatability, both in terms of the flesh of birds and their eggs. He was, unfortunately, a poor experimentalist, duped by his own enthusiasm. Even allowing for the difference in the way science was conducted then compared to now, and what constituted 'evidence' in the 1940s, Cott's research on the palatability of birds' flesh (not eggs) was flawed and later shown to be wrong. He also found conspicuous eggs to be less palatable than cryptically coloured ones. But this research was defective, too. His testing protocol involved presenting lightly cooked scrambled eggs of different species to a panel of human tasters. Cooked? What predators ever experienced cooked eggs? There were other methodological problems, including the fact that he regarded the eggs of all passerines – including the blue eggs of blackbirds – as cryptic. Overall, there is very little evidence that conspicuous eggshell colour signals distastefulness.[17]

Natural selection works in ways that can appear mysterious to us, especially if the target of selection is unclear. Suggestions about that target are limited only by our imaginations, and researchers – usually behavioural ecologists – who tackle problems like the adaptive significance of egg colour pride themselves on their imaginative hypotheses. They have generated three such hypotheses to account for conspicuously coloured eggs.

The first is referred to as the 'blackmail hypothesis' and proposes that brightly coloured eggs have evolved to coerce males into providing additional parental care – in the form of either incubating the eggs or feeding the incubating female – to protect the eggs from predators. The idea is that females evolve conspicuously coloured eggs that if left exposed attract predators and scupper the breeding attempt. To circumvent this males are forced to do some incubation themselves, or provide the incubating female with sufficient food so that she's less likely to forage for herself and leave the eggs unattended. Hmmm.

The second idea is that conspicuous egg colour reflects a female's quality and that the brighter her eggs the more her male partner is likely to invest in that female and her clutch. Specifically, the hypothesis is that a greater concentration of biliverdin – which is known to have antioxidant properties – provides a measure of the quality of both the female and possibly her offspring. As a result the more biliverdin the female can put into the egg colour, the more effort her partner is likely to provide. It has been known for a long time that chickens that are stressed, sick or both, produce eggs with less pigment, so the idea is not unreasonable.[18] Intriguingly, the pied flycatcher provides evidence for both aspects of this idea: better quality females – those in good body condition – do lay more intensely blue coloured eggs, and females that lay brighter eggs do get more help from their partner.[19] Other researchers have questioned the logic of the idea by asking how visible the eggs are in a cavity nesting species like the pied flycatcher?[20] On the other hand a study of American robins, an open-cup nester, in which natural clutches were swapped for artificial clutches that varied from pale to vivid blue, found that at nests with vivid blue eggs the male partner provided more food for the chicks than at nests with pale blue eggs.[21] While these two studies seem to provide clear evidence for this hypothesis, the jury is still out. More species need to be tested, and it is only when particular studies have been successfully replicated that scientists feel truly confident about the results.

Third is the idea that conspicuous eggs, and in particular white eggs laid in open nests on the ground, are adaptive because they provide protection from solar radiation and ultra-violet radiation. Experiments confirm that solar radiation at least can be an issue. A study conducted in the 1970s in which chicken and laughing gull eggs were painted either khaki or white and exposed to the sun for an afternoon revealed that the internal temperature of the khaki-painted eggs was 3°C higher than the white eggs. A similar experiment, in which naturally white or cream ostrich eggs were darkened with brown crayons and exposed to the Kenyan sun, gave essentially the same result: the darkened eggs were 3.6°C warmer than white eggs, and reached an average internal temperature of 43.4°C, which is beyond the lethal limit (42.2°C) for embryo survival.[22] One might ask therefore why laughing gull eggs are khaki and ostrich eggs white. The answer is that the laughing gulls breed in areas where daytime temperatures are never as high as those in Kenya. In addition, because they are susceptible to predation by crows and ravens the gulls rarely leave their eggs unattended, so they are rarely exposed to the sun. The ostriches on the other hand often leave their eggs unattended, but are vulnerable only to the depredations of Egyptian vultures which the adult ostriches can easily deter. A trade-off therefore exists between predation and thermal stress, which in the gull is tipped towards predation and hence favours a dark, cryptically coloured egg, whereas in the ostrich thermal stress is the greater risk and favours a pale, remarkably conspicuous egg.

Our second explanation for the colour of birds' eggs involves avoiding brood parasites. How do parasites like the common cuckoo manage to match their eggs so closely to those of most of their hosts? It was once thought that a female cuckoo looked at its host's eggs, captured an image of them in its brain, then sent that image to the uterus where it was reproduced on the surface of the egg she was about to lay. Colour photocopiers and scanners can do this, but it would be a tough call for a cuckoo. Even though some people thought that this was how the match was achieved, few egg collectors or ornithologists ever believed it. They knew that individual female common cuckoos always laid eggs of the same colour, which meant that they did not or could not adjust their egg colour.[23]

So how does matching occur? There are several possibilities, including the idea that cuckoos search until they find a host whose eggs are similar to their own.

To explore this idea, let's switch our attention from the European common cuckoo to another rather less familiar brood parasite, the cuckoo finch.

It is hot, around 40°C, and it is hard to breathe. The air is redolent with the scent of sunburnt grass. Earlier in the day I stepped off a plane from London and am now in Choma, Zambia, visiting my colleague Claire Spottiswoode who is studying brood parasitic birds. Our shopping complete, we drive through the shabby outskirts of

town on dirt roads and head into what to me seems like bush, but
to Zambians is farmland. After several miles we arrive at the farm of
Major John Colebrook-Robjent, an expat who grows tobacco. His
black workforce lives in tiny mud huts or in a few cases in home-
made brick huts, under minimal conditions, in an area known as
the 'compound'. I have been here before and met these men (but
not many of their wives) and each time I'm surprised by how much
they vary in manner and appearance. Some you'd be happy to have
round for dinner; others you'd give a wide berth to even during
daylight if you didn't know them. They are the nest finders. They
locate nests – extremely efficiently – either for the major or for
Claire, who has been studying the major's meticulously catalogued
collection of brood parasite eggs for her research.

The major died in 2008 aged seventy-two, and is survived by his
wife, Royce, who I haven't met before. She's in the garden when we
arrive, and from a distance I am shocked by how young she looks.
But it is an illusion: on closer inspection she is wearing a blonde
wig, one of a large collection of accoutrements. The next time I saw
her, without a wig, she looked her true age.

Growing up in London during the 1930s, like many of his
contemporaries the young Robjent was a keen egg collector. His
boyhood hobby metamorphosed into an obsession, and he later
became 'one of the most remarkable oologists of the twentieth-
century'.[24] On leaving school he joined the army, and after several
different postings was sent in 1963 to Africa. In 1969, however, he
resigned his commission and despite having no previous farming
experience settled in Zambia where he decided to try his hand at
growing tobacco.

Any interest in farming was quickly replaced by his passion for
birds and Robjent trained his farm boys to find nests, to skin birds
and to label specimens. One, who started at the age of ten, was
Lazaro Hamusikili, a man eventually so proficient he became both

indispensable and a liability. Lazaro's success was rewarded with cash, which, as he got older, he spent on alcohol, and as a result he would disappear for days at a time. When I first met him in 2008 Lazaro was in his early forties, and Robjent had just died.

Royce welcomed Claire and me into the house and allowed us to examine the cabinets containing her late husband's collections. The number of eggs was vast and, in contrast to Lupton's guillemot eggs that I later saw at the Natural History Museum in Tring, these were meticulously labelled.

Robjent is buried in the bush not far from the house but his reputation flares out behind him like the contradictory tail of a comet. In one light, his was a story of success; in another, he seemed to be little more than a common criminal. Which one you saw depended on your own perspective. Robjent's craving for eggs was widely known in ornithological and conservation circles. Indeed, as Robjent reported in his diary for 6 October 1988: 'In the morning I was "visited" by a team led by [the] Senior Assistant Commissioner of the Anti-Corruption Commission, plus four staff, also 2 Livingstone Museum people and 2 scouts of National Parks and Wildlife Service . . . Certain files, letters, registers, lists and diaries . . . were taken away.' Tipped off by evangelical conservationists, the Zambian Wildlife Authority (ZAWA) charged Robjent with possessing eggs without a licence or certificate of ownership. The trial took place in Choma on 22 December of that year. As Robjent said in his diary, the court was well attended by many local people and many friends, one of whom brought along a stack of cash in case of a possible fine – 'a splendid gesture'. Starting at 11 a.m., the trial was over by 12.30. After hearing from several people about the excellent work he was doing for Zambian ornithology, the magistrate 'told the court that he . . . considered my work to be of great benefit to ornithology and to Zambia and Zambians in particular, saying that I was doing useful work for the benefit of our children and our children's children . . . I was given an Absolute Discharge'. The magistrate also told the authorities that Robjent 'should be given every encouragement . . . with the issue of collecting permits'.

For whatever reason, the permits never appeared.[25]

Robjent's meticulously catalogued collection of eggs and skins was also well known to the Natural History Museum in the UK, and after his death they made arrangements for it to be transferred to Tring. One could be forgiven for thinking that this might be a smooth operation involving little more than some minor paperwork greased by the goodwill between the authorities in Zambia and its colonial parent, from whom it had become independent in 1964. Not so. The failure of ZAWA to issue the necessary permits to Robjent after his trial proved to be a major stumbling block that has resulted – so far – in an unresolved delay in the move to the UK. Luckily, soon after Robjent's death, Douglas Russell, the curator of eggs at Tring, went to Zambia, catalogued the collection, packed it up and moved it to another farm for safe-keeping. It is fortunate he did, for a couple of years later when I returned to Zambia with Claire, Royce had died, and although the house was locked and ostensibly guarded, it had been ransacked. Whatever one might feel about egg collecting, it seemed tragic that Robjent's notebooks and many of his other belongings might be either lost or scattered among the filth on the floor of what was once his home. An important piece of history had been laid to waste – equivalent to losing the connection between eggs and data I'd seen in Lupton's collection of guillemot eggs.[26]

Robjent's special interest was brood parasites and their hosts, in particular the cuckoo finch, whose primary hosts are the tawny-flanked prinia, the red-faced cisticola and a handful of other nondescript little brown birds. Also known as the parasitic weaver, the cuckoo finch male is a lovely saffron yellow, whereas the female is an undistinguished mottled brown, like most female weaverbirds. Appropriately, its scientific name, *Anomalospiza*, means 'strange finch'.[27]

What appealed to Robjent about the cuckoo finch was not only the extraordinary variation in both its eggs and those of its tawny-flanked prinia host, but also the often incredible match between the eggs of the two species when they were found in the same nest. The matches between the eggs of brood parasites and their hosts have enthralled generations of egg collectors. Perhaps the most famous (or infamous) of these was Edgar Chance, a wealthy English businessman who in the early part of the twentieth century amassed around 1,600 common cuckoo eggs and the clutches of their accompanying host eggs. Although ostracised by the ornithological community for his dodgy dealings over illegal eggs, Chance made some astute observations and produced two important books contributing substantially to our understanding of the common cuckoo's biology.[28]

Robjent knew from his collection that there must be different 'types' of female cuckoo finch, each producing a certain egg type, and each laying only in the nests of a specific host, whether it be the prinia or the red-faced cisticola.[29]

He never got round to writing anything about the cuckoo finch, but, before he died, Claire visited him. On seeing his collection of eggs she could see that the cuckoo finch must be much more abundant than she imagined, and was – with the help of Robjent's nest finders – perfectly feasible to study. She was hooked, and over the next few years worked on the major's farm, employed his nest finders and made some remarkable discoveries.

When a host like the prinia is parasitised by a cuckoo finch it wastes a lot of time and energy rearing the young of the cuckoo finch and none of its own. Cuckoo finch females typically lay an egg in a prinia nest when the host has still to complete its clutch of two, three or four eggs. The cuckoo finch egg hatches a day or two before those of its host, and its chick monopolises the food brought by its foster parents so that when the host chicks hatch they quickly die of starvation. From an evolutionary perspective, a prinia parasitised by a cuckoo finch is a dead loss – literally. It is hardly surprising, then, that if a host were able to identify and eject a cuckoo finch egg it would have greater breeding success than those lacking this ability. However, the more that hosts like prinias learn to recognise and reject parasite eggs, the greater the selection pressure on the cuckoo finch to produce an egg that the prinia cannot distinguish from its own. Egg recognition and ejection by hosts has led to egg mimicry by brood parasites like the cuckoo finch.

It is an arms race. The better the hosts become at recognition and ejection, the greater the pressure on cuckoo finches to produce an egg that will fool the host. One outcome of this arms race is for the host to produce more and more diverse eggs, which is exactly what the prinias have done. The eggs of different females vary in colour, from those whose ground colour is a pale blue, olive brown to foxy red, and they can be marked with khaki, black or red blotches or squiggles.

As Claire began her fieldwork, evidence for the arms race between cuckoo finches and prinias emerged. She noticed that many of the cuckoo finch eggs she was finding had a blue ground colour, whereas most of those in Robjent's cabinets, collected during the 1980s and from exactly the same location, were predominantly red. It looked as though a change had occurred.

Claire's careful analysis showed that not only had the colours of the cuckoo finch eggs changed over just forty years, but so had the hosts' eggs; they had done so almost in parallel. It appeared that as the hosts evolved more and more extreme egg colours, the cuckoo finches had tracked these changes in time. As Claire said: 'This makes sense if the dominant colours and patterns of prinia eggs serves as a good defence only until cuckoo finch mimicry catches up. At that point, natural selection should begin to favour prinias that lay eggs of novel appearance which make the cuckoo finch eggs stand out.'[30] It would be hard to find a better example of a co-evolutionary arms race between host and parasite.

In colour and pattern the prinia's extraordinarily variable eggs are like miniature guillemot eggs. In both species this variation is the result of intense selection – and in the case of the prinia by brood parasitism. In the guillemot's case it is individual recognition.

I love Wallace's idea that the bright colour and elaborate patterning of guillemot eggs might be due to this species' natural exuberance. Guillemots are indeed exuberant birds. It is one reason why I like them so much, but, sadly, this isn't why they lay such wonderful eggs.

It seems to have been Thomas Pennant in the late 1700s who first suggested that the enormous variability in colour and pattern might allow guillemots to identify their own egg. It was an assumption, of course, and Pennant provides no context for his suggestion.[31] A century later, in 1878, the banker and amateur ornithologist John Gurney, writing about the egg collecting at Flamborough, records how he was told something similar by a Mr Leng who 'affirms that they know their own eggs, and I [Gurney] believe: for what purpose can so great a variety of markings be given to them?'[32]

A more complete statement occurs in Henry Dresser's magnificent multi-volume *Birds of Europe*, published in the 1880s, where he writes that Mr Gray 'who in his pleasing account of Ailsa Craig . . . ingeniously suggests that, "where such large numbers of eggs are crowded together the great diversity of marks and colouring may enable each bird to distinguish its own" but then says that in rainy weather the eggs become dirty and hence doubts it'.[33] Gray's statement is that the variation in egg colour is an adaptation to high-density breeding because it allows guillemots to recognise and care for their own eggs.

Science proceeds by testing hypotheses and often by conducting experiments that may or may not provide evidence that supports the hypothesis. It was with the explicit purpose of conducting experiments that Beat Tschanz, whom we have already met, set out in the late 1950s to test this hypothesis.

How would *you* go about this? One way would be to climb on to a guillemot breeding ledge, scaring away all the birds as you did so, swap a guillemot's egg for another of a different colour and see if it was accepted by the returning parent. In fact, if you do this, as Les Tuck – a Canadian guillemot enthusiast employed by the Canadian Wildlife Service – did on Funk Island in the 1950s, you would find that: 'When birds returned to their nest sites, they invariably continued to incubate in the same location, failing to react adversely to the substitute for their own egg.' On the basis of his simple experiments Tuck concluded that the variation in colour was nothing to do with recognition but, instead decided that colour was camouflage.[34]

Tuck's experiment was too simple. Designing a good – that is, biologically appropriate – experiment depends on knowing something about the birds' biology and some common sense. Imagine you are a guillemot. Someone comes along and scares you off your egg. You fly off out to sea in a huge loop before returning to your breeding site where you know you left your egg. There's an egg present, but it doesn't match the one you left. You have two choices: accept it and resume incubation, or reject it. If you choose 'reject', you are likely to have zero breeding success that year. If you accept, then because there's a chance that it might be your egg – albeit slightly different in appearance – you have greater than zero breeding success. After all, the egg was in the right place, and after all eggs sometimes get splashed with dirt and change their appearance, so – on balance – the best option might be to accept.

Now think about designing a better experiment that avoids the shortcomings of Tuck's. The obvious way of doing this, which tests whether guillemots can distinguish their own egg from that of a neighbour's, is to give them a choice, or, rather, to force them to make a choice. Because the breeding site, an area half the size of your hand, is so important, you would need to think of a way of separating the bird's preference for its site from its preference for its egg. You could do this by recreating a biologically realistic situation and moving the egg a few centimetres away from the breeding site, as though it had rolled slightly, but also placing another egg the same distance away, so that when the guillemot comes back it has to decide which of the two eggs to retrieve and start incubating.

This is essentially what Tschanz did: and, yes, guillemots do recognise and retrieve their own egg. His studies also suggested that they *learn* the colour of their egg. Tschanz found that soon after laying, he could substitute a guillemot's egg with another one of a different appearance that the parents would accept, presumably because they were still learning the colour and pattern. Later in incubation the birds were much more reluctant to accept a strange egg. However, if egg colour were changed gradually, equivalent to the problem birds have as their egg becomes soiled by dirt and faeces, the guillemots would accept a different coloured egg. Given that females produce a very similar egg throughout their lives I wonder to what extent they can remember from year to year what their egg looks like. Male guillemots face a slightly different problem since they must learn to recognise their partner's egg, and they probably have to be more flexible in this respect, because over their lifetime they might breed with more than one female.

Tschanz also checked to see whether guillemots used shape – which is also variable between, but consistent within, females – to help them identify their eggs. But there was no evidence for this; as long as an object was of an appropriate size and was painted the 'right' colour and pattern, the birds would retrieve and incubate it, regardless of whether it was a cube, a rectangular block or a truncated pyramid.[35]

In the late 1980s and early 1990s Tony Gaston, who was working at Coats Island in northern Hudson Bay, decided to repeat Tschanz's (1959) egg-recognition experiments on Brünnich's guillemots. Both guillemot species show considerable variation in egg colour and patterning, and both breed at high densities, although Brünnich's guillemots never breed on broad, flat areas or at such high densities as common guillemots. What Gaston and his colleagues found was that when given a choice of two eggs, one their own, one a foreign egg a few centimetres away from the true breeding site, the birds usually identified their own egg and then either sat on it and began incubating, but then shuffled the egg back to its original site, or remained standing and, using their bill and feet, manipulated their egg back to its correct site. Clear evidence, then, that Brünnich's guillemots, like common guillemots, recognise their own egg.

However, Gaston and his colleagues also noted that Brünnich's guillemots were more likely to accept a foreign egg than a common guillemot, possibly because the broader ledges that common guillemots breed on provides them with more incentive to look harder and further afield for their egg following a disturbance. Common guillemots have been seen retrieving their egg from as far as four metres from its original site, whereas for a Brünnich's guillemot the maximum distance they are ever likely to be from their egg is a few centimetres along a ledge. The fact that Brünnich's guillemots are slightly less discriminating in their egg recognition may account for why, when returning to their site and finding their own egg missing, they sometimes steal a neighbour's egg – something not known to occur in common guillemots.[36]

There are few other birds that breed in such close proximity – and without a nest – as the two guillemot species. Some terns, such as the royal tern and Caspian tern, breed close together but with distinct nest sites, and they can distinguish their own egg from a foreign egg only if the difference is particularly marked. Also, the markings on the eggs of these species seem to have evolved primarily for camouflage rather than to facilitate individual recognition.

It was once thought that the small birds that play host to the common cuckoo also recognise their own egg – how else were they able to throw out a cuckoo's egg, which they often do? However, some of the first research – and it was extremely limited – suggested that their ability to identify a cuckoo egg was not based on knowledge of what their own egg looked like, but simply on the fact that the cuckoo's egg was different from their own. This understanding was based on an experiment conducted by Bernhard Rensch in the 1920s in which he replaced the eggs of a garden warbler with those of a lesser whitethroat as soon as the garden warbler's eggs were laid. When the garden warbler laid its fourth egg – which Rensch left in the nest – the garden warbler removed it. Rensch assumed that the ejection occurred because the fourth egg was different from the others in the nest, but it is just as likely that the parent birds had learned what (they thought) their own eggs looked like in the previous few days. Indeed, subsequent research on gray catbirds, which are parasitised by the brown-headed cowbird in North America, showed that learning is a crucial part of egg recognition and avoiding the eggs of parasites.[37]

Cuckoos and cowbirds create one kind of problem: they are interspecific brood parasites, laying their eggs in the nests of species other than their own. More insidious perhaps and even harder to detect is intraspecific brood parasitism, parasitism by your own species. This is known to occur in several birds, including common starlings, certain weaverbirds, coots and moorhens. It is easy to see how being able to tell you've been parasitised and being able to do something about it would be advantageous. Of course from the parasite's point of view this is the last thing you want – so just as with the prinia and the cuckoo finch, there's an arms race between hosts and parasites.

One of the best-studied intraspecific brood parasites is the American coot. In the British Columbia population investigated by Bruce Lyon of the University of California at Santa Cruz, an average of 13 per cent of all eggs in a nest are from parasitic females. Since the chicks of parasitic coots survive at the expense of the host's chicks, parasites have a substantial negative impact on breeding success and we would expect coots to have evolved ways of minimising these costs. Coot eggs have a highly variable background colour and markings, and host coots use this variation to identify foreign eggs, which they simply bury in the nest material. The fact that eggs disposed of in this way are more dissimilar from the host's eggs than those that are not buried, strongly suggests that recognition is based on these egg characteristics.[38] Lyon later went on to ingeniously test that rejection was based on true recognition rather than on a mere difference, as Rensch suggested for his garden warbler. By swapping eggs between nests he created clutches in which the proportion of host and parasitic eggs was equal, and in eight out of the twelve experimental clutches the parent coots chose to bury or push to the periphery the parasite eggs rather than their own eggs – and in none of these twelve nests were any host eggs treated in this way.[39]

In fact, all the evidence – separate studies of seventeen different species – shows that true recognition is how birds identify their own eggs. Presumably true recognition involves simply learning the appearance of one's own eggs, whereas identification based on differences requires recognition and – in the case of birds laying several eggs – the ability to assess the proportion of eggs of the two types.

True recognition also occurs in the ostrich, where several females lay in a single nest and the enormous clutch is incubated mainly by a single male. The most senior female remains with the male and the rest go off – presumably in search of another nest to dump some more eggs into. The senior female's aim is to ensure that as many of her eggs as possible are properly incubated, so she pushes the eggs of the other females either to the periphery of the nest or out of it altogether. She seems to be able to do this by recognising the pattern of pore openings on the creamy white, unmarked egg surface.[40]

The study of egg colouration has come a long way since Alfred Russel Wallace's Darwinian speculations. Like many other areas of biology, the investigation of egg colour has been boosted by technological innovations. In recent decades researchers have had the benefit of digital cameras that allow us to measure and quantify colour both relatively easily and more precisely than previously. We also have a much better appreciation of the way birds see the world – including their eggs – and that birds can see into the ultraviolet part of the spectrum and distinguish more colours than we can. These developments have promoted what some of the key researchers in this area have called a 'renaissance in the study of eggshell pigmentation'. As well as providing answers to some important questions, these studies have increased our appreciation of the extraordinary diversity of eggshell colour and pattern – created essentially through the use of just two pigments. Finally, like all good research, this renaissance has raised a host of new questions.[41]

From the colourful exterior of eggs we move next to the colourless region of the interior – the albumen.

6

Much Ado About Albumen:
The Microbe War

The egg-white varies in quantity and consistency in different groups of birds . . . unlike the yolk it is completely used up before the bird hatches.

O. Heinroth, *Aus dem Leben der Vögel* (1938)

There is a distinct 'nothingness' about albumen. It is the most unobtrusive part of a bird's egg – and for several reasons. For a start, it is colourless, even though its name derives from *albus*, meaning white, which is, of course, what it becomes only after its proteins have been de-natured and transformed by cooking. Second, albumen seems to be structureless: a poorly defined glob of mucous-like material. Third, as a child, I was told that the albumen is 90 per cent water, immediately reinforcing the notion of nothingness: if it's mostly water then it can't be up to much.

The truth is that, far from being unimportant, albumen is absolutely remarkable, mysterious stuff. Its role in the developing egg is vital, providing water and proteins for the growing embryo and at the same time cushioning the embryo from physical damage as the egg is turned or rolled around inside the nest. But much more crucially, the albumen provides a sophisticated biochemical firewall against the microbes that, given half a chance, would consume the developing embryo.

This chapter looks at the remarkable succession of barriers – of which the albumen is the most important – that have evolved to defend against the microbial invasion of birds' eggs. Since several different mechanisms work together to this end we are going to jump about a bit, so be prepared.

We'll start by looking at how and where the albumen is made. For Greek scholars such as Aristotle and those of the early Renaissance like Fabricius and William Harvey, the fact that an egg was composed of two very different parts – albumen and yolk – was a puzzle. If the developing embryo needed nourishment, why wasn't a single substance sufficient? More mystifying still, where did the albumen come from? Aristotle's account starts by dismissing the belief that the two parts come from the two sexes: yolk from the female and white from the male. He reaffirms that both come from the female, and in a roundabout way suggests that the yolk itself produces the albumen.[1] Some two thousand years later Fabricius has a much more accurate account:

> Whilst the yolk is rolling down and slowly rotating through the . . . uterus, it gradually also collects from thence a part of the white which is there generated and prepared to be laid down around the yolk, until, having passed beyond the twists and turns in the middle and being arrived at the last one, the yolk along with the white of the egg encompassing it, is wrapped up in membranes and acquires a shell.[2]

Unexpectedly, Harvey writes: 'Being taught by experience, I incline more to Aristotle's opinion'[3] but it is far from clear why he thinks that.

Harvey also criticises Fabricius for writing the following:

There is another use of the albumen when it is segregated from the yolk, namely that the foetus may swim in it and so be supported lest it should sink down of its own weight . . . And to this the viscidity and purity of the albumen contributes. For if the foetus remained in the yolk, it would easily settle down into its depths and even rupture the yolk.[4]

This is almost too much for Harvey who says: 'All very unsatisfactory! For what in heaven's name has the purity of the albumen to do with supporting the foetus? Or how can the white which is thinner more easily support it than the yolk which is thicker and grosser?' He adds that the foetus 'does not swim within the albumen or in the yolk, but in the fluid that I have called the colliquament [the fluid beneath where the embryo starts to develop, technically the blastoderm]'.[5]

The discovery that albumen is formed in that region of the oviduct called the magnum happened in the mid-1800s, as zoologists scrutinised the world through rapidly improving microscopes.[6] Later, we'll see just what albumen comprises.

When I was a child, my mum regularly served a favourite supper of soft-boiled egg with 'soldiers'. On one occasion when I was very hungry I whipped the top off the egg, plunged my spoon inside and took a mouthful. No sooner had the egg entered my mouth than I spat the whole lot out across the table in a state of absolute revulsion. The egg was bad, off, rotten or whatever you want to call it. I have never tasted anything so revolting before or since.

At the time my mother was fairly blasé about my teatime experience because, as I later learned, this was a period when the long-term storage of eggs by farms and shops was routine – sometimes for as long as a year – and addled eggs were not uncommon. She told me that my rotten egg was probably a forgotten egg; one that had been neglected long enough for its contents to start to decompose.

That egg played havoc with my taste buds for only a few seconds but fixed the memory in my brain for ever. This was an egg whose microbial defence system had failed. The characteristic smell (and taste) of rotten egg is hydrogen sulphide, a by-product of microbial activity. Perhaps we shouldn't be surprised that commercially produced eggs are occasionally invaded by microbes; after all, we have little idea of the conditions under which those eggs are collected, stored and transported. Was my rotten egg the result of an infected hen? Was it because the shell was cracked during transport, allowing the microbes to get inside? Or perhaps it was intact, as I seem to remember it was, but had sat for weeks or months in physical contact with an infected egg?

Well aware of the risk of microbial infection, commercial egg producers in the 1960s washed eggs in an attempt to remove any bugs from the shell before they went to market. Ironically, washing only makes things worse for it removes the cuticle, the egg's outermost covering whose specific role is to keep microbes out of the pores, and if the temperature of the water is cooler than that of the egg, microbes are pulled into the pores as the egg dries. Rather than hampering microbes, washing actually helps them gain access.[7] The problem was eventually solved by a more sophisticated system of washing and drying. That was in America. The European Union forbids the washing of eggs, arguing that overall unwashed eggs are safer, and some statistics I found on the web seem to support this.

Contamination with chicken faeces after laying and before washing makes the risk of microbial infection even greater. For many years one of the technical staff in our department kept chickens at home and brought the eggs into work to sell. They weren't cheap but they were popular – because he told everyone that they were 'organic', an idea he reinforced by deliberately adding an occasional feather to a carton or by smearing one of the eggs with a bit of chicken faeces. In a candid moment he once told me that most of the eggs he sold to us he had bought cheaply from a supermarket. It is ironic that that smear of chicken faeces that made the eggs so appealing could also have made us ill.

The microbes that can infect eggs include bacteria, viruses, yeasts and fungi. I was lucky to suffer no more than gustatory revulsion and a tainted memory, for the commonest bacterium found in hens' eggs is salmonella, a potential killer.[8] It is precisely because of this health risk that we know as much as we do about the defence mechanisms of eggs.

The risk posed by microbes in eggs was brought into sharp focus in 1988 when a junior health minister for the Conservative government in the United Kingdom, Edwina Currie, announced that 'most egg production [in the UK] sadly is now infected with Salmonella'. Her comment caused chaos: egg sales and consumption collapsed and millions of hens had to be slaughtered. What Currie should have said was that most commercial *flocks* of hens were infected with salmonella, and not most birds. It was an expensive ambiguity, both for the British egg industry, and also for the government, which had to pay out huge sums in compensation to the egg producers. It was costly for Currie, too, since she was forced to resign.

Interestingly, the source of infection of the eggs was internal, that is, some birds within the commercial flocks had salmonella in their bodies, which they then transferred to their eggs. Years later it was revealed that Currie was right: there had indeed been a salmonella epidemic in eggs in the UK during the late 1980s, but the government had covered it up.[9]

The UK poultry industry subsequently invested millions of pounds in trying to produce safe, bug-free eggs by developing a salmonella vaccine. Fortunately for those of us in the UK, they largely succeeded. In 1997, for example, before widespread vaccination there were 14,700 cases of salmonella poisoning in the UK, but by 2009 this had fallen to around 600. There is still a risk today from imported eggs, whose hens have not been inoculated. And in the USA there are over 100,000 cases of salmonella poisoning each year – because, it is argued, the cost of vaccination is too high.[10]

What about wild birds? Are their eggs susceptible to microbial invasion? Unless infection is exclusively a consequence of the mass production and processing of hens' eggs, it seems likely that they are. What is remarkable is that even though microbial infection of hens' eggs, via the pores, had been demonstrated as early as 1851 it wasn't until the early 2000s that we knew that the eggs of wild birds were also susceptible to microbial infection.[11] We've known for much longer that bacteria and other microbes are everywhere and thrive wherever there's a good food supply – and that includes the tissues of almost any other living organism. Animals like us, and indeed adult birds, protect themselves from microbial attack through an immune system. An egg is a rich source of nutrients but has no immune system. Birds have circumvented this particular problem by creating eggs with multiple layers of defence in which the albumen plays a central role. No biological system is perfect, however, and, as soon as an egg evolves a way of keeping bugs at bay, then there is pressure on the microbes to find another way into an egg. It is a bug and bird arms race.

Even though it passes through a bird's bottom there are surprisingly few bacteria on the surface of an egg when it is laid. There are bacteria in birds' nests so the shell soon acquires bacteria. And while those on the egg surface cannot do much harm, if they can penetrate the egg either through the pores, or through any hairline crack in the shell, the embryo inside may be doomed.

One of the first to realise that the pores in the eggshell provide an open door to microbes was John Davy, whom we met earlier, and who in the 1860s described how on opening an egg that had been twenty-two days under a hen: 'I found it's [sic] inner membrane covered in part with mould (*Mucor mucedo*) the spores of which must have entered, it may be presumed through the foramina [pores] in the crust [shell].'[12]

The person who confirmed that the outermost layer of the shell is the egg's main way of keeping microbes from entering the pores was Ron Board in the 1970s. An undergraduate at the University of Bristol in the 1950s, Board completed a PhD at Edinburgh on bacterial infection of the hen's egg and then obtained a post as lecturer at the newly formed University of Bath. Introducing himself as 'Board by name but not by nature', he considered himself a 'food safety microbiologist'. This doesn't sound overly exciting, but in fact he was very much a zoologist, and enjoyed keeping ornamental poultry and waterfowl – often the surplus from his research – in his garden. In many ways Ron Board epitomised the successful academic of the 1960s and 1970s. Sporting a tweed jacket and puffing studiously on his pipe, he enjoyed the scientific freedom and easy funding that was peculiar to that golden era, which in his case resulted in some fundamental discoveries about the eggs of wild birds.[13] The turning point in Board's career occurred in the summer of 1973 when he examined some eggshell surfaces with a scanning electron microscope.

Invented in the 1930s, it wasn't until the mid-1960s that the scanning electron microscope (SEM) became commercially available and a routine (if extremely expensive) piece of kit in university science and engineering departments. The SEM produces seductively beautiful and revealing 3-D images, and Board was sufficiently seduced to change the direction of his research.

Before describing Board's findings, we need a bit of background about the egg's surface. A study in the 1840s showed that the outermost covering of the hen's egg is composed of organic material now referred to as the cuticle.[14] A similar organic, protein-based cuticle occurs on the eggs of other birds such as tinamous, kiwis and jacanas. In contrast, the surface of the eggs of seabirds, including gannets, shags and pelicans, but also the guira cuckoo, flamingos and grebes, is inorganic and comprises calcium salts.[15] In fact, there are two types of calcium: a soluble form of calcium carbonate known as vaterite and the more durable calcium phosphate. Because the outermost surface of the egg can be either organic (as a cuticle) or inorganic (as calcium salts), it was decided to use the term shell accessory material (SAM) to cover both types. Regardless of whether this outer covering is organic or inorganic, the thickness of the SAM layer, which varies from 30µm in the anhinga to 60µm in the gannet, constitutes about 11 per cent of total shell thickness.[16]

Board examined the eggshells of a wide range of birds and noticed that those species that laid and incubated their eggs in wet or muddy locations – such as grebes and flamingos – had an unusual surface. The immense power of the SEM revealed that the surface of these birds' eggs was covered with a mass of tiny spheres, each one measuring only half of one micron in diameter.

It was already known that the eggs of various water birds have an unusual bloom – sometimes powdery, sometimes chalky and sometimes waxy – and in the 1960s one of my ornithological heroes, David Lack, director of the Edward Grey Institute for Field Ornithology in Oxford, suggested that these unusual features were for waterproofing. However, as Board pointed out, Lack failed to say *why* waterproofing might be important.[17]

In fact, David Lack must have been aware that the embryo inside would not be able to breathe if the pores in the shell were covered by, or full of, water. What he could not have known was how the waterproofing was achieved, for that was what Board discovered.

Board wrote:

There appears to be no record of waterlogging of shells leading to the asphyxiation of embryos, but the experiences of the
poultry industry leave no doubt that the flooding of a few
pores with contaminated water is the first step in the process
leading to the addling of eggs.[18]

Here's the hint that it was not simply asphyxiation that was potentially dangerous for water birds: microbial contamination was also
a problem. In one of his first papers Board spelt out the issues:
eggs *can* be contaminated by bacteria and their route in is through
the egg's pores. Micro-organisms are most effectively transported
in water so waterproofing the eggshell is what keeps microbes at
bay. The waterproofing has to be done in a way that still allows
the embryo inside the egg to breathe – rather like a biological
Gore-Tex – and, as Board discovered, this is achieved by covering the pores with shell accessory material, and most effectively
by a layer of microscopic spheres which, as we saw in Chapter 2,
creates a surface that by its physical nature repels water and hence
is unwettable.

Board was also aware that eggs have a second line of defence
beyond the cuticle. This is the inner shell membrane, whose ultra-
fine, mesh-like structure is thought to act like a net to trap bacteria.[19]
The inner shell membrane is not an infallible barrier, because some
types of bacteria can digest the fibres, effectively creating a hole
through which they gain access to the next layer inside the egg, the
albumen – a nice (or not so nice) example of the ongoing arms race
between microbes and eggs.

Since the early 1900s it has been known that certain bacteria
(species of *Bacillus*) will not grow on egg albumen, amply demonstrated by the fact that if you leave a blob of albumen in a dish for

two or three months it remains free of bacteria. This was the first
indication that there was something exceptional about albumen
and it was Alexander Fleming – of penicillin fame – who in 1922
identified what this was.[20] Albumen contains a bacteria-destroy-
ing protein, which Fleming called lysozyme. He gave it this name
because 'lysis' means to break down (in this case, the cell walls of
bacteria), and 'zyme' refers to it being an enzyme, and means to
ferment. Lysozyme has since been found in our tears, in saliva and
some other body fluids where its antiseptic properties are important.[21] But lysozyme is only one of several antimicrobial proteins
in albumen. By the 1940s no fewer than five proteins capable of
inhibiting microbial growth had been identified. By 1989 this
number had risen to thirteen; and then, with the advent of new
technologies such as proteomics in the 2000s, over one hundred
antimicrobial proteins in albumen have been identified and it
seems likely many more remain to be discovered.[22]

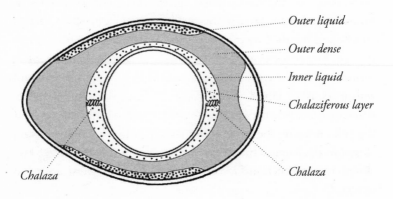

The four types of albumen and their location in a bird's egg.
Redrawn from Romanoff and Romanoff, 1949.

Let's take a closer look at albumen. Superficially, there doesn't appear to be much to see, but by cracking a really fresh hen's egg on to a plate it is immediately obvious that albumen is not homogeneous. There are no fewer than four types of albumen, and in four concentric layers. If we can get technical just for a moment, working from the outside in, there is an outermost liquid layer (23 per cent of the total albumen volume), followed by a dense, viscous layer (53 per cent), an inner liquid layer (17 per cent) and finally, next to and surrounding the yolk, a denser chalaziferous layer (3 per cent). This latter layer, secreted by the infundibulum, probably also forms the fine filaments at opposite sides of the ovum, which when twisted with other filaments become the two chalazae. These are the white, amorphous stringy bits we see when we crack an egg into a bowl, and are most frequently encountered as the glutinous knobbly bits when we eat scrambled egg. The chalazae are created as the ovum spirals through the magnum region of the oviduct. Their role is to suspend the yolk within the albumen and this works because one end of each chalaza is attached to the ovum itself (via the chalaziferous albumen layer) and the other is firmly lodged in the layer of dense viscous albumen, which itself is attached to the shell membranes at the pointed and blunt ends of the egg. The chalazae also allow the yolk to rotate when the egg is turned so that the embryo always remains on top of the yolk and within the inner liquid albumen layer.[23] This 'self-righting' ability is achieved because the embryo develops on the least dense side of the yolk. Keeping the embryo uppermost ensures that it is always closest to the parent's brood patch for maximum warmth, but also closest to the inner surface of the shell for maximum access to oxygen.[24]

Perhaps the most surprising mechanism by which the albumen keeps microbes at bay is that albumen contains nothing that a microbe needs or can exploit: the albumen's 'nothingness' has a purpose. Albumen contains few nutrients that would support microbial life and those nutrients that are present are locked up by certain types of protein that make them inaccessible to bacteria. To a microbe, the journey across the albumen from the shell membrane to the yolk on the inside is equivalent to a human trying to walk across the Atacama Desert: there's nothing there to sustain life. It would be hard to think of a more economical way to keep bacteria and fungi at bay. It is also intriguing to think that the amount of albumen in eggs, and hence the size of eggs, might be driven in part at least by the need to keep microbes away from the yolk.

If this is hard to imagine, try this: lay a hen's egg on its side and with sharp-pointed scissors cut and remove a circular window of shell a couple of centimetres across. Now look into the egg: this is its real structure, the centrally lying yolk suspended by the milky chalazae and surrounded by the glaucous albumen. Now imagine you are a salmonella microbe, just 2μm long, that's just burrowed its way through a pore and the double shell membrane, but like someone standing on the threshold of a desert, is deterred by the prospect of trying to traverse this succession of four vast, lifeless zones.

But the albumen has two more defence strategies up its sleeve. The first is its alkaline nature (a pH of 9 or 10) – a fact first noticed by John Davy in 1863 – which microbes generally don't like. Second, many of the antimicrobial proteins in albumen are thought to be most effective when they are relatively warm – specifically at the temperature at which the egg is incubated.[25]

If we want to know whether microbes in eggs are likely to be a problem for birds, and if we want to understand how birds have overcome the risk of microbial infection, we should look at those species breeding in moist and dirty conditions where microbes thrive.

To me this suggests three types of birds: those in the humid tropics; mound builders – chicken-like birds that bury their eggs in heaps of rotting vegetation instead of incubating them using their own body heat; and guillemots, which often incubate their single egg in a guano quagmire.

The study of microbes in birds' eggs is a relatively new field of research and as recently as 2004 it was not known for certain whether microbial infection of the eggs of *wild* birds could kill embryos.[26]

The importance of microbes in the eggs of wild birds was discovered by accident. Steve Beissinger, a biologist and ornithologist at Berkeley, University of California, was interested in the question of why certain tropical birds – in contrast to many temperate birds – begin incubating so soon after their eggs are laid. Birds lay only one egg each day and since some species produce clutches of ten eggs, if incubation were to start only when the clutch was complete – as it does in many temperate breeding birds – the first laid egg would sit around for ten days before being incubated.[27] The low ambient temperatures experienced by birds in temperate regions keeps the embryo within the egg in a state of suspended animation for two or more weeks. In the tropics, however, the warmer ambient temperatures preclude this cool state of suspended animation. Beissinger and his colleagues decided to test the idea that the immediate onset of incubation in tropical birds occurs because their eggs have a shorter shelf life than those in temperate regions.

Working on the Caribbean island of Puerto Rico, Beissinger's study species was the wonderfully named pearly-eyed thrasher – a common thrush-like bird. The idea was to move freshly laid thrasher eggs from the warm, dry lowlands to the cool, humid uplands – up and down an altitudinal gradient – and hold them there for a week before putting them back under the females to incubate. The prediction was that eggs that had been kept cool at the higher altitude would have a longer unincubated shelf life.

It turns out that the shelf life of thrasher eggs was extremely limited, with hatching failure rising from 21 per cent for eggs held for one day to 98 per cent for those held for seven days. But, significantly, there was no difference between the lowland and upland site. Clearly, temperature was not the most important factor. However, what Beissinger and his colleagues found was that many of the eggs held at the cool, humid high-altitude site were infected by a very visible fungal growth that killed the embryo.

This unexpected discovery made Beissinger think. He knew that poultry biologists had already spent a lot of time looking at the effects of microbes on eggs and he wondered whether their findings might help him understand what was happening in his thrashers. In fact thirty years earlier Ron Board had anticipated Beissinger's approach: 'Comparative studies of shells . . . and of albumen . . . of eggs of other species suggest that the principles derived from poultry research may well have general application.'[28]

Under natural conditions pearly-eyed thrashers start to incubate as soon as their first egg is laid. The warmth caused by the incubating parent is critical because it increases the temperature of the egg to the point where the antimicrobial enzymes in the albumen are thought to be most effective. Together with the fact that incubation also keeps the egg dry, this is what controls the number of microbes. Indeed, incubation is so effective that in Beissinger's study no naturally incubated eggs became infected.[29]

It is obvious from this that if the thrashers followed the same strategy as some temperate breeding birds and deferred incubation until the clutch was complete, many more eggs would fail as a result of microbial infection. To check that was correct, Beissinger and his colleagues conducted another study of similar design but this time on temperate breeding birds, and, as predicted, found no increase in

microbial infection when eggs were left unincubated. Birds breeding in the tropics therefore seem to be more susceptible to microbial
infection of their eggs.[30]

At the very dawn of scientific enquiry in the 1660s, Francis Willughby
and John Ray decided that their *Ornithology* (see Chapter 1) should
include a description and illustration of every known bird. At the
time there were thought to be about five hundred species (there were
actually around ten thousand), but this was still incredibly ambitious because so little was known, especially about birds outside
Western Europe. For these Willughby and Ray often had to rely on
the accounts of travellers and explorers, and in many instances they
found it difficult to distinguish fact from fantasy. What to accept
as true and what to reject as nonsense? To avoid throwing the ornithological babies out with the bath water completely they created
an appendix in their encyclopaedia of 'such birds as we suspect for
fabulous'. One of the birds in their appendix is a species they called
the 'daie' of which they wrote:

> It is very strange (more strange I dare say than true) that so
> little a bird should lay so great eggs, and so many together,
> and in such deep vaults under ground, and being there hidden
> they should be hatched without being ever sitten upon or
> cherished by the old ones, and that the young once hatched
> should of themselves presently fly away.[31]

Ray concludes by saying: 'I dare boldly say that this history [account]
is altogether false and fabulous.'

It wasn't. The 'daie' is the Philippine megapode, and that
description – which came originally from Antonio Pigafetta who

accompanied Magellan on his 1519–21 expedition to the East Indies and the Philippines – is pretty close to the remarkable truth. The Philippine megapode is one of a group of large-footed birds from the Indo-Pacific region and Australasia that hatch their eggs by burying them either in warm volcanic soil or in piles of fermenting vegetation.[32]

The two best-known megapode species are the malleefowl and the brushturkey, both of which are found in Australia and both of which arrange for their eggs to be incubated in piles of rotting vegetation. It is hardly surprising that such an unusual system of incubation has resulted in some extraordinary adaptations. Not only are these piles of decomposing vegetation extremely humid, they are full of microbes, and of course it is the micro-organisms that generate the heat on which the eggs' development depends.

In the early 1980s, suspecting that they might be unusual, Board examined the eggs of the malleefowl. Almost as predicted, the outermost covering of their eggs comprises a vast number of minute spheres that he assumed keep both water and microbes out of the pores. In addition, the spheres contain very little organic material which may prevent them being corroded by micro-organisms.[33] Thirty years later, just after Board's death, another research group examined the eggs of the Australian brushturkey, which also have an outer covering composed of numerous tiny spheres of calcium phosphate.[34]

It isn't just birds' eggs that are vulnerable to microbial infection; crocodile and alligator eggs, like those of mound-building birds, are also incubated in piles of rotting vegetation. Although hardly subject to much research, it is known that the albumen in the eggs of these reptiles also has antimicrobial properties.[35] This may be a very ancient trait because the albumens of the eggs of snails, fish and frogs are all able to resist microbial infection.[36]

It has been known since the 1940s that the way albumen deals with microbes varies considerably in different bird species[37] and the researchers studying the brushturkey eggs expected them to have albumen that was especially resistant to microbes. But as far as they could tell, their albumen was no more effective against microbes than that in hens' eggs. I have to admit to being surprised by this result, because however effective the outermost covering, nature usually puts several failsafes in place. Imagine a mutant microbe that found a way of circumventing the brushturkey's eggshell defence system. With no antimicrobial proteins in the albumen there would be nothing to stop it, once inside the egg, from reaching the yolk and eating the embryo. The authors offer an explanation for this somewhat counter-intuitive finding, suggesting that the addition of substances to the albumen to make it more effective against microbes may have a negative effect on the developing embryo.

That idea is essentially the same as one proposed in the 1970s by American biologists Gordon Orians and Dan Janzen in a paper provocatively entitled 'Why are embryos so tasty?' Their idea was that eggs that are so attractive and palatable to predators would be better off if they contained something that – as in certain caterpillars – made them distasteful. The fact that no birds' eggs are actually distasteful – eat your heart out Hugh Cott (see Chapter 4) – is explained by the suggestion that anything unpleasant in the egg would slow down the embryo's growth rate and be more disadvantageous than beneficial.[38]

There is at least one bird that is thought to deliberately render its eggs unpalatable. Occurring across much of Asia, Mediterranean Europe and southern Africa, the hoopoe has been described as looking like a tiger-banded ice axe. With striking cinnamon, black and white striped plumage, a sickle-shaped beak and a beautiful crest, this is indeed an attractive and unusual bird. It is probably because of the contrast between its appearance and its disgusting domestic habits that the hoopoe has had a prominent place in mythology. Aristotle commented that it makes its nest of dung, especially 'man's dung'. Much later, the French naturalist the Comte de Buffon wrote: 'It has been long said and often repeated, that the hoopoe besmears her nest with the excrements of the wolf, of the fox . . . of the cow and all sorts of animals, not excepting man; and that she does this with the view to defend her young by the loathsome stench.'[39] Buffon then says that while they don't actually plaster their nest with dung, 'the nest is indeed very dirty and offensive [because] the young ones cannot throw out their excrements and therefore grovel a long time among filth'. He adds that this is undoubtedly the reason for the saying 'As nasty as a hoopoe'.

Buffon is correct: hoopoe nest cavities are filthy because the chicks' very liquid droppings cannot be removed by the parents. In many other birds nestlings produce their faeces neatly wrapped in a gelatinous coating that allows the parents to pick them up and dispose of them outside the nest. Hoopoe chicks deter predators trying to get into the nest by squirting them with particularly obnoxious liquid faeces.

An additional source of unpleasant odour in the hoopoe nest is the preen gland of the female and the chicks. The fact that the male hoopoe's preen gland secretion is unexceptional, while that of the female and her nestlings is disgusting, immediately suggests that there is something interesting going on. And there is. Discovered and reported by Christian Ludwig Nitzsch in 1840, the malodorous brown and oily preen gland secretion from female and nestling hoopoes was initially thought to be solely a deterrent to nest predators.[40] Over a century later, however, the secretion was found by Spanish researcher Juan Soler and his colleagues to contain symbiotic bacteria that protected the birds' plumage from a particular feather-degrading bacterium.[41] The antimicrobial bacteria in the hoopoe's preen gland are 'good bacteria', a bit like the so-called good bacteria in certain yoghurts, only much less pleasant to us. When the Spanish researchers deactivated the good bacteria in hoopoe nests, they found that fewer of the birds' eggs hatched, suggesting a mutually beneficial arrangement between the birds and the bacteria over and above any positive effect on the birds' plumage.

The researchers then looked at the way preen gland secretion enhanced breeding success. Their unexpected discovery was that hoopoe eggs have no outer covering; no shell accessory material. This is surprising because, as we've seen in other birds, the outermost layer of the egg is the first line of defence against pathogens. In fact, the hoopoe's unusual egg surface had been described in the mid-1800s by Wilhelm von Nathusius (who we met earlier) and who, as well as commenting on the lack of cuticle, also noticed that: 'sunk into [the eggshell] surface, are open pits [that] stand so thickly together . . . they look like perforations in a sieve'. Nathusius's interest in eggshells was almost entirely descriptive and he rarely dared to speculate why certain eggshells – of any species, including those of the hoopoe – were the way they are.[42] But that's exactly what the Spanish researchers did after noticing that, within a few days of incubation, the tiny pits in the eggs' surface became filled with material – including the good bacteria – from the preen gland. Wondering whether it was the presence of preen gland bacteria on the eggshell that controlled pathogenic microbes, the researchers conducted a clever experiment where, by placing a temporary cover over the gland, they prevented some female hoopoes from depositing preen gland material on to their eggs. Females treated in this way hatched fewer eggs than those allowed to anoint their eggs with preen gland material: QED.

The hoopoe example is striking in several respects. First, just as in other birds, antimicrobials are important, but rather than being present in the albumen they are derived from the preen gland and applied externally to the egg.[43] Second, the surface of the hoopoe's egg, with its lack of any outer covering and numerous tiny pits, appears to have evolved specifically to hold the preen gland secretion. This seems odd to me, for the following reason. Imagine a proto-hoopoe whose eggshell has no cuticle, that starts nesting in cavities that become filthy and full of harmful bacteria. Natural selection would favour any mutant that could protect its eggs from dangerous microbes. The simplest mutation would result in a hoopoe that produced eggs with a protective cuticle. Much more complex would be a succession of mutations that produced, first, an unusual preen gland secretion with antimicrobial properties; second, an unusual eggshell surface with pits; and finally, a special behaviour in which the female anoints her eggs with preen gland secretion.

Although the hoopoe case is slightly puzzling, there's a hint that preen gland secretions keep egg-surface microbes under control in other birds, too. A comparative study of 132 bird species found that the relative size of the preen gland was positively associated with the total surface area of the clutch – which is exactly what you might expect if it is important for birds to ensure that each egg acquires a covering of preen gland secretion.[44]

And what of guillemots? In my mind's eye I can see George Lupton at the end of each day returning to his Bridlington boarding house to examine his haul of eggs. I can also sense his euphoria as he holds that season's Metland egg in his hands and feasts his eyes on its colourful surface. Judging from the muck on the outside, the egg has been incubated for a couple of days, but Lupton wipes away any trace of the egg's previous life with a damp cloth. By arrangement with his landlady, he's at the kitchen table with an empty dish and a bowl full of water and his egg drills laid out like a surgeon's implements in front of him. Each drill is a fluted, cone-shaped metal contraption, rather like an elongated version of the device joiners use to countersink screws. Picking up the first egg, Lupton looks for a point at which he can penetrate the egg's surface. As the spike bites, and as Lupton twizzles the drill gently between his finger and thumb, there's a soft fall of snow-like calcium carbonate on to the table. Lupton pushes, turning the drill harder until he feels it 'give' as he breaks through the inner shell membrane and the drill becomes wet with albumen.

He twists the drill back and forth a few times to produce a perfectly symmetrical circular hole about six millimetres in diameter. Every collector has his own way of doing this and perfection is part of the process. Lupton then takes his brass blowing tube, bent almost at right angles, and, placing the narrow end just at the entrance of the hole he's made, he puts the other wider end into his mouth. Positioning the egg over the empty dish with the hole in the egg facing downwards, he proceeds to blow. As anyone knows who has done this, blowing out the contents of an egg takes some effort. With Louis Armstrong cheeks and bulging eyes, Lupton forces the ice blue albumen from the hole. The different fractions of albumen come in fits and starts: the outermost liquid fraction runs out easily, almost like water; the dense viscous fraction resists until the last moment and then splutters out like a noisy fart, but some of it sticks and Lupton has to use his fingers to pull it free. With a final exhalation of relief the globules that contain the chalazae emerge. Then, slowly and serenely comes the yolk: a continuous, opaque golden stream running smoothly into the bowl. Almost the last to appear is the germinal disc, a tiny speck among the yolk whose red threads are the beginnings of the 'placental' blood vessels. Seeing this confirms what Lupton already suspected: this was an egg laid almost as soon as the previous collection was complete, for it has clearly been incubated for two or three days. A fresher egg would have had no sign of development.

Lupton finishes by flushing the interior of the shell several times with water. After they've drained for a few minutes he collects all of the eggs he's blown and takes them up to his room to dry overnight – safe from accidents. As soon as he's gone the landlady swoops in to take the bowl of yolks and albumen that Lupton's left – by arrangement – and scrambles them for her son's supper. Waste not, want not. Throughout the six-week-long guillemot season at Bempton, the climmers collected eggs every three days. Their aim was to ensure that the eggs were fresh enough for human consumption, conveniently for the collectors who needed to relieve the eggs of the contents prior to adding them to their display cabinets. Blowing a fresh egg is infinitely easier than emptying one that contains a developing embryo. Another fortunate feature of collecting guillemot eggs at three-day intervals was that most would be relatively clean, with insufficient time to be too badly soiled by the guillemot faeces that cover the breeding ledges. If the weather is dry, guillemot ledges are dry, too. But after rain, guillemot ledges quickly degenerate into the fishy equivalent of a pig farm and their eggs rapidly become smeared with guillemot shit. After I've been handling dirty eggs or climbing around on wet guillemot ledges it takes two or three days to get rid of the smell no matter how many times I wash my hands.

Of a sample of 112 eggs that I examined (on a dry day), midway through incubation on Skomer in 2014, *all* had some shit on the shell, and on average 10 per cent of the shell surface was covered, but some of them had more than half their shell covered and two eggs were completely covered. The smell indicates that guillemot shit is the kind of muck microbes would find attractive – fish-based, and alternately warmed and dried on the ledges. It wasn't until I started this study that I realised that every guillemot egg you see in a museum collection or on the pages of a book is immaculate and unnaturally clean.[45] If guillemot eggs are always smeared to a greater or lesser extent with shit, how do they cope with such an execrable incubation environment? How do they avoid their eggs being penetrated by the many microbes there must be on the ledges?

The answer is that we don't know because no one has looked. Indeed, until I started writing this book I hadn't really appreciated the full impact that shit has played in the evolution of the guillemot's breeding biology.

On the basis of what we know about mound builders and hoopoes, there are several ways guillemots could cope with a filthy environment. One possibility relates to the fact that the pigment protoporphyrin – responsible for the dark markings on most birds' eggs, including those of the guillemot – provides some protection from microbes.[46] However, a study by Phil Cassey and his colleagues found that the concentration of protoporphyrin on guillemot eggs is not particularly high,[47] and anyway the amount of dark markings is so variable between guillemot eggs I think we can safely discount this.

Another possibility concerns the eggshell surface, which in other birds incubating in damp, dirty conditions consists of nano-spheres that might keep water and microbes out of the pores. The surface of guillemot eggshells consists of a layer of minute pimples covered with nano-spheres that some researchers think is hydrophobic. If it is, it must work in a different way from that of the brushturkey and, so far, it is not at all clear how this provides any protection from the bacteria in guillemot faeces. We started off using scanning electron microscopy to examine the surface structure of the guillemot's eggshell, but soon found an even better, more revealing technique: X-ray micro-computer tomography (micro-CT for short). This is the same technology as that used in hospitals, but on a microscopic scale. It shows in extraordinary detail the remarkable difference in the shell surface of the guillemot and razorbill egg – a difference whose significance we don't yet understand, but which is surely related to the issue of dirt, for, as we've seen, guillemot eggs are invariably contaminated with faeces, but razorbill eggs are not.

Moving inwards, we get to the egg membrane, which could be important in this respect for it is particularly robust in guillemots. Next is the albumen, but we still have to ascertain whether guillemot albumen has especially effective antimicrobial properties.[48]

We need to compare our dirty incubators – the pearly-eyed thrasher, the brushturkey and guillemot – with some birds whose eggs are incubated in super-clean environments. What is super-clean? Dry, I think. Several birds whose eggs we know about fall into this category: they include the woodpigeon, the ostrich and several small passerines such as warblers and flycatchers. None of these species have any cuticle on their eggshell, presumably because in the dry environment in which their eggs are incubated the risk of microbial infection is so small – although, to be honest, we don't know.[49]

Guillemots are not quite unique in incubating shit-covered eggs. Certain ducks deliberately defecate on their eggs when scared off their nest by a human or a fox. The canvasback, tufted duck, shoveller and common eider all do this, and the idea is that their foul-smelling faeces deters predators from consuming the eggs. One observer writing in the early 1900s described the eider's 'green and oily excrement' as being totally different from ordinary excrement . . . and of such a frightful odour as to 'deter the hungriest dog'.[50] Experiments confirmed that crows preferred not to eat eggs contaminated with eider faeces. This is fine, providing the faeces on the egg surface have no detrimental effect on the developing embryo, either by blocking the pores of the egg so the embryo cannot breathe or by encouraging microbial infection of the egg. Does anyone know? Do the shells, albumen or membranes of the

A common guillemot egg. The egg has been photographed on a breeding ledge and has a few flecks of fecal material on its surface; museum specimens, in contrast, are always immaculately cleaned.

FLAMBOROUGH EGG GATHERING. A FIND.   596

Above: A 'climmer'
descending the Bempton
cliffs in the early 1900s.
Above right: George Lupton
(known as 'FG' [Frederick
George] to his family and
friends) in the 1920s or
1930s.

Right: Lupton's ten-year-
old daughter Patricia
emerges over the clifftop
carrying two guillemot
eggs that she has
collected, 21 June 1931.

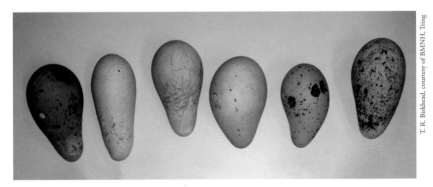

T. R. Birkhead, courtesy of BMNH, Tring

Variation in the shape of birds' eggs. *Top* (*from left to right*): common crane, kittiwake, great cormorant, great cormorant, peregrine falcon and great northern diver; *middle*: grey heron, hen harrier, hobby, European starling, missel thrush, carrion crow, golden plover and osprey; *bottom*: these are all aberrantly shaped guillemot eggs collected by George Lupton at Bempton.

American robin.

Great tinamou.

Wattled jacana.

Striated heron.

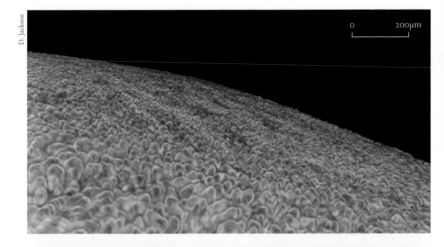

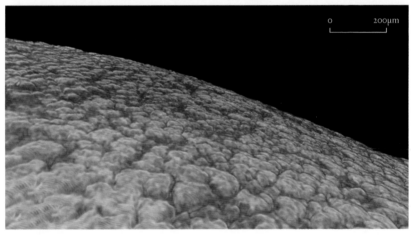

Micro-CT scans of the eggshell surface of a common guillemot (*top*) and a razorbill (*bottom*). Notice the 'pimply' surface of the guillemot egg. The scale (200 μm = ⅕th mm) relates only to the foreground in each image.

A guillemot egg in white vinegar as the calcium of the shell dissolves away (*top left*) and after all the calcium has gone and the 'egg' is held only by its shell membranes, but still bearing the colour (*top right*). Below, two examples of sections through the yolk of guillemot eggs show the daily growth rings.

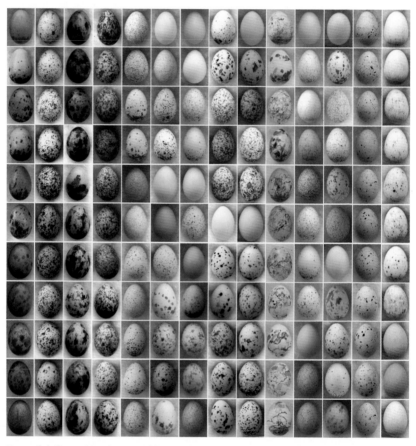

Eggs of different Zambian warblers and weaver birds (each column is a different species) that are parasitised by diederik cuckoos and cuckoo finches. The hosts have evolved signature-like variation in their egg colours and patterns, both within and between species, to help them detect the parasites' eggs which mimic their own. This allows them to recognise and reject them from their nests.

eggs of these particular ducks possess any special features that ameliorate the effect of their mother's faeces?

Often in nature similar problems are solved in broadly similar ways. But adaptation isn't perfect because there are constraints. Sometimes these are dictated by the birds' evolutionary history, and natural selection has to work with what it has got – how else are we to explain the different antimicrobial strategies we have just looked at? How else do we explain the hoopoe? Of course, there could also be factors that we haven't even thought about.

This chapter has been concerned mainly with the wonderful antimicrobial properties of albumen and we will finish with a further celebration of its significance.

Somerset Maugham's compelling semi-autobiographical novel *Of Human Bondage* reinforced the deeply entrenched idea that the only worthwhile part of an egg is the yolk. The story describes how an aunt and uncle, the Reverend William Carey, bring up the orphaned, club-footed Philip Carey. It is a loveless existence and on one occasion Philip is told to sit still and stay quiet while his uncle has a nap. Protesting that he cannot sit still until teatime, Philip is instructed by his uncle to learn by heart a prayer; his uncle adds, 'If you can say it without a mistake when I come in to tea you shall have the top of my egg'. There is a double poignancy here, for, apart from the pointlessness of learning the prayer, the top of a boiled egg – consisting only of the white – is in most people's minds a long way from being a treat.

Philip Carey might have been less disappointed had he known that, despite its apparent nothingness, the albumen in a bird's egg is at least as important as the yolk – and in at least three ways. First, it is albumen that separates reptiles from birds. Reptile eggs contain very little albumen; instead the water they need for embryo development comes from the environment, soaking in through the leathery shell from the soil or vegetation in which their eggs are laid. Because birds' eggs are warmed by a parent and usually in a nest this isn't an option, so from the moment they are laid bird's eggs must contain all the water the embryo needs. This water is in the albumen. Second, and confirming what I have just said, experiments in which different amounts of yolk and albumen were removed from hens' eggs revealed that reducing the amount of yolk had little effect on the chick other than it having a smaller reserve of yolk inside its body at hatching. When albumen was removed from the egg, however, the embryo's growth was retarded and the chicks that hatched were stunted. It was initially assumed that this was because removing albumen denied the chicks some critical proteins, but subsequent studies revealed that the effect was to deprive the chicks of water that was essential for their growth and development.[51] And third, as eggs get bigger, by which I mean eggs laid by bigger species of birds, all components increase in volume, but the albumen increases disproportionately. Larger eggs, produced by larger species, tend to have relatively more albumen than they do yolk.[52]

Having established the vital role the albumen plays in the egg, we next consider the yolk – effectively the food-laden, female sex cell.

7

Yolk, Ovaries and Fertilisation

All animals come of eggs.
John Ray, *The Ornithology of Francis Willughby* (1678)

Of all mythological symbols the serpent is among the most powerful, simultaneously representing fertility, rebirth and sexual desire, and as one of my colleagues recently wrote, 'the snake is the midwife to the birth of the human condition'.[1] Although its elongate body form – so suggestive of both the penis and spermatozoa – has such obvious male connotations, it was actually female snakes that provided some of the most extraordinary initial insights concerning the reproduction of both birds and humans.

Willughby and Ray's *Ornithology* included a brief overview of what was then known about 'generation' – the term used to encompass both reproduction and embryo development. Although unaware of the existence of sperm, Willughby and Ray were, through their numerous dissections, well aware of the essential role of the testes and ovaries. In particular they comment on the appearance of the bird's single (left) ovary, a feature that had been noted by Aristotle, and whose structure in birds is so different from that in mammals. A bird's ovary in the breeding season resembles a bunch of grapes and is 'made up of a large number of oocytes [ova] held together by connective tissue, joined to a heavy stalk by which they are held in place in the body'.[2] Outside the breeding season, however, the oviduct is reduced to a thread and the ovary resembles a collection of millet seeds – so much so that Willughby and Ray refer to them as 'seed eggs', a name that simultaneously alludes to their appearance and their potential. Come the breeding season a handful of those 'seeds' expand and are filled with yellow yolk. Even to the uninitiated the inference is obvious: the large ova are mature; the tiny white seed eggs are immature. With a bit of practice you can even tell how many ova have already been released from the ovary that breeding season, for each one is released from the equivalent of a little bag (a follicle) that persists in the ovary, even when empty, and is then referred to as post-ovulatory follicle. As Ray and Willughby comment, the number of ova in a bird's ovary – some maturing, most simply waiting – is so enormous that it looks like it could easily be a lifetime's supply:

> It is most probable, that hen-birds have within them from their first formation all the eggs, they shall afterwards lay throughout their whole lifes time; so that when their cluster of eggs is wholly spent, they cease breeding, and become effete [i.e. exhausted]: as Angelus Abbaticus has observed in vipers . . . Now why should nature prepare so great a stock of yolks (which as we said would suffice for many years births) if she had given to females a faculty of generating new ones? Neither is it true only of birds, but also of all female quadrupeds, yea, and of women themselves, that they have in them from the beginning, the eggs or seeds of all the conceptions, they shall afterwards bring forth through their whole lives.[3]

There's quite a lot of information embodied in this paragraph, so we need to dissect it carefully so as not to miss anything. Willughby and Ray's objective is to tells us about reproduction in birds, but to do that they have borrowed an idea from a sixteenth-century physician, Baldus Angelus Abatticus (usually referred to simply as Abati), whose main claim to fame was an excellent little book on vipers published in 1589.[4]

Interested mainly in their venom, Abati dissected vipers and described their internal structure, including that of their reproductive system. He noted that in some females the ovaries contained a 'huge supply of eggs' (he means ova) while others had none at all. The reason for Abati's surprise was that several earlier 'authorities', notably Pliny (writing between AD 23 and 79) and Galen (writing a bit later: AD 129–200), had claimed that vipers reproduce only once and then die.[5] If that was true Abati must have thought, why do some females contain so many ova – many more than the number of eggs they were known to lay? Ingeniously, Abati infers that the female vipers with abundant ova are young individuals, while those with none are old and have exhausted their supply of ova. In other words, Abati says, both Pliny and Galen were wrong in thinking that vipers reproduce only once. Such a challenge to the authority of the ancient authors marks the very beginning of the scientific revolution. Indeed, so confident was Abati about his findings that he entitled Chapter 9 of his book (albeit somewhat clumsily): 'Why it may not be true that the viper reproduces only once; but reproduces many times is established' (sic).

Willughby had a copy of Abati's viper book in his library, but one has to read both Abati and the *Ornithology* carefully to figure out Willughby and Ray's train of thought.

Essentially, Willughby and Ray use Abati to infer that birds are like vipers, starting life with a full set of ova, which when used up are not replaced. However, their statement that the same is true in quadrupeds and 'yea, women themselves' is sheer speculation. It appears that they inferred a rule: vipers, birds. . . and, yes, mammals, too, are all the same. We know that their comment about women must have been a wild extrapolation because the human ovum was not seen until 1827 and it was not possible to count the ova (oocytes) accurately within a human ovary – which, as we'll see, was crucial for this argument – until the 1950s.[6]

The fact that between puberty and middle age a woman's supply of eggs (ova, oocytes) is both limited and rapidly diminishing is considered 'one of the most basic doctrines in the field of reproductive biology'.[7] The concern that many women feel about the ticking of their biological clock is hardly surprising for the decline in the number of ova – from half a million to almost zero in fifty years – is exponential.

If you were to ask human fertility researchers who first came up with the notion that a woman's supply of eggs is set at birth, they would probably name Heinrich Wilhelm Gottfried von Waldeyer. Born in 1836, Waldeyer became one of Germany's top anatomists, and one who believed that the study of embryology – whose heyday was around this time – was fundamental to the study of anatomy. Recognising that there was little existing knowledge on how the reproductive organs take form during embryo development, he made this the focus of his research, studying a wide range of organisms, including birds. The year 1870 saw the publication of his masterly textbook *Eierstock und Ei* ('Ovary and Egg') in which – unaware of Abati, Willughby or Ray – he claims to have discovered that both birds and girls start life with their full complement of eggs.[8]

For fifty years Waldeyer's view of a fixed number of ova dominated reproductive biology. But then two American researchers, Edgar Allen in 1923 and Herbert Evans in 1931, independently argued that on the basis of what they considered 'incontrovertible histological evidence', the process of making ova was continuous. Their idea was that 'a wave of atresia (destruction of the population of reproductive cells) was followed by a wave of regeneration in which thousands more oocytes were formed'.[9] Waldeyer's idea was pushed off its pedestal.

Solly Zuckerman (later Lord Zuckerman), well known for his pioneering work on the reproductive biology of monkeys and apes in the 1920s, was one of those persuaded by the Americans' view. He described how he rejected Waldeyer's idea in favour of Evans and Allen's new doctrine. There was a compelling logic to the idea that women continue to make ova throughout their reproductive life (i.e. until middle age), just as men continue to produce sperm throughout their (longer) reproductive life.[10]

During his research, Zuckerman and his colleague Dr Anita Mandl discovered a way of counting oocytes – a notoriously tricky procedure – in the ovaries of rats and began to doubt the Americans' view, wondering whether Waldeyer might be right after all. Through their study Zuckerman and Mandl found very clear evidence that the number of oocytes decreased exponentially with age. They offered two possible explanations. Either there is a finite supply, as Waldeyer proposed, or, if ova continue to be produced, the rate of production decreases rapidly with age. Zuckerman favoured the first idea and by the early 1950s was confident he had enough data to reinstate Waldeyer's hypothesis.[11]

We now know that a female human foetus aged about eighteen to twenty weeks has about 300,000 non-growing follicles (oocytes – potential eggs) in each ovary; that this has fallen to 180,000 at age thirteen years, to 65,000 by age twenty-five, 16,000 by age thirty-five and at the age of fifty there are no more than one thousand left in each ovary.[12]

These are huge numbers, so what is the problem? The maximum number of children ever born to one woman is sixty-nine (from a succession of multiple births), so even at the age of fifty, with a thousand oocytes left in each ovary, there seems to be more than enough. Why women and other female vertebrates produce such an excess of ova and store them for so long is a mystery. The huge numbers of sperm produced by males is much more easily explained,[13] but, crucially, sperm aren't all produced at birth and instead are created continuously, removing any concern about their shelf life. The puzzle is why a system in which a female's ova are stored for so long evolved. The vast numbers of ova make sense in that this allows for a large proportion of ova not surviving long enough and in good enough condition to be used to start a new life. We also know – at least we think we know – that the defective ova are selectively removed and eliminated during menstruation.[14]

The idea of a fixed number of ova at birth has recently been challenged once again – this time by researchers exploring the role of stem cells, and partly on the basis of logic – again. Wouldn't it be more sensible, the proponents of this view ask, to make ova as they are needed, ensuring they are fresh and healthy?[15] Other researchers don't give this 'new idea' much credence and have dismissed it as a publicity gimmick designed to persuade women desperate for a child to pay for expensive and unproven fertility treatment. The consensus is that women do *not* make new ova.

Let's return to birds. Willughby and Ray's motivation for writing about birds (and they planned other books on fishes, insects and plants)[16] was to provide an objective account of what was actually known. This was the beginning of scientific knowledge and it meant evaluating everything that had previously been written about birds; discarding old wives' tales and retaining information they felt they could trust or had verified for themselves. This wasn't quite as onerous a task as it sounds, for Willughby, a wealthy aristocrat, had a considerable library and, compared with what we know today, ornithological knowledge in the seventeenth century was limited.

But it was their approach that differed from what had gone before: they were objective. As part of this new 'scientific method'

Willughby and Ray were also meticulous in stating where they obtained their information, allowing their readers to know what they had discovered themselves and what was derived from previous writers. Scientific honesty was an innovation, as most contemporary writers were rampant plagiarists, assuming ownership of all previous knowledge, passing it off unacknowledged (and usually uncritically) as their own. For their material on reproduction Willughby and Ray relied to a large extent on William Harvey's book *Disputations Touching the Generation of Animals*, which after a long gestation had been published in 1651. While they were clearly in awe of the great man, they were not afraid to criticise him when they felt he was wrong. Intriguingly, Harvey seems to have been unaware of, or disagreed with, Abati's idea that females begin life with a fixed supply of eggs. Indeed, it appears that Harvey thought otherwise and referring to the existence of the tiny ova in the hen's ovary he says: 'For although the hen has as yet no rudiments of eggs ready prepared in her vitellary [ovary], nevertheless being made fertile by coition, she shortly produces eggs in her ovary.'[17]

In response, Ray writes in the *Ornithology*: 'I am not ignorant that Dr Harvey . . . does assert, that though a hen has no seed-eggs within her, yet after coition she will breed new ones' . . . 'But I think that Great Naturalist did not sufficiently consider or examine this matter.'[18] In other words, Ray thinks that Harvey overlooked the numerous, undeveloped ova in the bird's ovary.

We now know that the ovary of a domestic hen contains several million ova, of which several thousand are visible to the naked eye. We also know, thanks to the extraordinary pioneering work of Raymond Pearl and William Schoppe in the 1920s, that, contrary to what Willughby and Ray – and everyone else subsequently – thought, birds *can* produce new ova. Pearl and Schoppe conducted experiments in which they surgically removed parts of the hen's ovary, and discovered that the number of ova subsequently returned

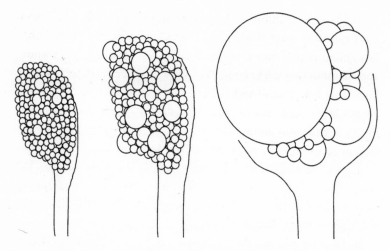

A typical bird's ovary at different stages of development. Left: outside the
breeding season with ova resembling millet seeds; middle: start of the
breeding season, some ova starting to swell; right: a mature ovary with a
hierarchy of yolk-filled ova.

to what it was prior to the procedure. It is very strange that – for
reasons we don't yet understand – birds have the ability to produce
new ova, but women don't.[19]

During the heyday of egg collecting in the eighteenth and nineteenth centuries it was commonly assumed that it took only a
day for a guillemot to produce an egg. The evidence seemed clear:
you climbed down on to a ledge and removed all the eggs, and if
you returned the next day, hey presto! There were more eggs. The
impression was that if a female guillemot was deprived of its egg,
she simply produced another one overnight.

The Bempton climmers knew this wasn't right. From the 1800s (and probably long before, but it was never written down) they recognised that if they removed a guillemot's egg, a second would appear about two weeks later. They were certain of this because they knew that female guillemots always laid the same coloured egg and always at the same site on the cliff ledge. So well known was this aspect of guillemot biology that Lupton and other collectors specialised in securing two, sometimes three, and once in a blue moon, four, successive eggs from the same female in a single season. The fruits of the climmers' and collectors' labours lie in groups of two, three or four in the glass-covered cabinets in the Natural History Museum at Tring, and elsewhere.

In fact, we can use the production of a replacement guillemot egg to establish just how an egg is formed. The single most important aspect of this is the production of the yolk: the true ovum – the single food-rich cell on which lies the female genetic material in the germinal disc. Creating the genetic material isn't a big deal, but accumulating the nutrients that comprise the yolk, and that will help to sustain the embryo as it grows, is what takes time.

Although a guillemot's egg is much larger than a hen's egg, the yolk (which is also larger and weighs about 35g) makes up about the same percentage by weight (31 per cent) as in a hen's egg. And, like the yolk of a hen's egg, it is full of nutrients, mainly fat and protein, which come from the bird's diet. Indeed, during the time that the yolk is forming – and remember that for the guillemot, as with most seabirds, it is only a single egg – the female stays out at sea, away from the colony, foraging to accumulate enough of these nutrients to make a yolk.

The trick to knowing what goes on during the period of egg formation was discovered in the 1890s by Luigi Daddi, an Italian scientist who fed chickens with a red dye and then examined the yolks of the eggs they subsequently laid. After hard-boiling the egg, Daddi cut the now solid yolk in half to reveal the layer in which the dye was deposited, confirming that, rather like an onion, the yolk was formed in successive layers. This must have been gratifying news for the British physician Allen Thomson, who, forty years earlier, in 1859, speculated that this was how the yolk formed. In fact, the idea that the yolk was fashioned in this way had been noted by Harvey in the 1600s but was probably over-looked or forgotten.[20] Harvey was trying to sort out whether the albumen is created by the yolk itself or by the uterus, as his tutor Fabricius suggested. Harvey favoured the former and suggested that Fabricius had been deluded into adopting his explanation because 'the white of an egg, being boiled hard, is easily split into layers lying one upon the other. But this also happens in a yolk that is still clinging to the ovary, if it be boiled hard.'[21]

In the early 1900s Daddi's use of dye to understand yolk forma-tion was exploited by other researchers, including Oscar Riddle who was later celebrated for discovering the pituitary hormone, prolac-tin. A few years later, in 1908, another researcher, Claude Rogers, noticed that however much dye he fed his hens their yolks were never uniformly coloured; they always consisted of alternating bands of red-dyed yolk and undyed yellow yolk. His results were consistent with Riddle's suggestion that the darker bands were created during the daylight hours when the hens were actively feeding, and the paler bands at night when they weren't. In other words, pairs of bands, one dark, one light, represented the daily deposition of yolk into the developing ovum. By measuring the thickness of the bands, Rogers was able to show that deposition was slow initially, and then speeded up, and that it took a hen about fourteen days to produce an entire yolk.[22]

Duping domesticated hens into swallowing capsules of coloured dye proved to be a good way to study yolk development, but persuading wild birds to do the same is more difficult. The

American poultry scientist Dick Grau took up the challenge in the 1970s. Based at Davis, California, Grau was unusual among poultry researchers in also being interested in wild birds. He knew that there were only a few species – notably seabirds – that could be easily caught and dosed with dye. It worked, and for both Cassin's auklet in California and Fiordland crested penguins, which Grau studied while on sabbatical in New Zealand, he was able to establish the way that yolk accumulated in the ovum.[23]

But there are many bird species where feeding a dye capsule is impractical. Another approach was needed. Several earlier researchers had noticed that, even without feeding dye, it was sometimes possible to detect faint alternating light and dark rings in the yolks of hard-boiled eggs. Grau reasoned that if he could find a way to render these rings more distinct, he would have solved the problem. After some trial and effort he found that freezing the yolks, slicing them in half and placing them face down in a solution of potassium dichromate for the best part of a day enhanced the yolk rings and created a beautiful onion-like pattern of alternating light and dark grey-green bands.[24]

Collecting fresh eggs is infinitely easier than catching female birds and dosing them with dye, and once Grau had his new technique up and running, so was he. With his colleagues, Grau found that the duration of yolk formation varied from a fast and furious four to five days in the red phalarope, to twelve days in the glaucous gull, twelve to eighteen in the common guillemot, and thirty days in the southern royal albatross.[25] In the 1980s, my colleagues and I also measured the rate of yolk formation in guillemots, confirming the values obtained by Grau's team, but also showing that replacement eggs had a shorter duration of yolk growth (on average 9.3 days) than first eggs (11.5 days), as expected, because females are under pressure to lay another egg as soon as possible once they've lost their first egg.[26]

Of the thousands of tiny ova in a guillemot's ovary, only a few – as we have seen – ever get the signal to go forward to the next stage and become charged with yolk. We do not know for any species of animal how that choice is made. It is hard to imagine that it is random. It is also hard to imagine that all ova are created equal, otherwise why start with so many? It is widely assumed that the ova that are chosen to receive yolk are of better quality, but again we have no idea in what way.

The selected ova start to receive the necessary nutrients to form yolk through the ovary's well-developed system of blood vessels that carry fats and proteins manufactured in the liver, as well as vitamins, minerals and pigments, to the ovary. This nutritious mix provides almost everything the embryo needs. Of the rest, the albumen provides water and protein, while the shell provides some of the calcium for the growing embryo's skeleton.

There must be a programme that dictates when the yolk forma-tion process stops. How many layers? How large a yolk? It wouldn't be too difficult to imagine what this control process might be if it was identical for all females of a particular species, but, as our guillemot results showed, it isn't.

Why is there so much variation in the size of yolk? There are several possible reasons. First, females in good condition may be able to forage particularly efficiently or have sufficient reserves to be able to produce relatively large yolks and give their offspring a better start in life. Conversely, females that struggle to find sufficient food may produce an egg as soon as they meet some minimum threshold to produce an adequate yolk that will grow an embryo to term.

A third possibility is that females strategically adjust the amount of yolk they put into their eggs to maximise the success of the offspring. This idea emerged from a study of captive zebra finches undertaken by Nancy Burley, a biologist based at the University of California, Irvine, in the 1980s. Burley made the extraordinary discovery that the plastic colour rings (bands in North America) placed on the birds' legs in order to identify individual birds – standard practice among many scientific ornithologists – altered their attractiveness to members of the opposite sex. Red rings made male zebra finches more attractive but green rings had the opposite effect. An explanation for this was that male zebra finches have a red beak, and the addition of red rings simply enhances this effect. Since green is not part of the zebra finch's plumage, green rings seem to make males unattractive.

After discovering this Burley realised that she could use colour rings to manipulate male attractiveness and get a better handle on how sexual selection works. Amazingly, she found that males with red rings lived longer and had higher reproductive success than green-ringed males – even though in captivity all birds were given equal access to food and water. One other signifi-cant finding was that females paired to red-ringed (attractive) males – that is, males they perceived to be of high quality – were prepared to work harder in reproduction than females paired to green-ringed (unattractive) males. This difference in the amount of effort allocated to rearing offspring she called 'differential allocation'.[27]

So far, none of this is about eggs, because Burley didn't look at the eggs of female zebra finches paired to males with different coloured rings. The point of mentioning it is to introduce the idea that female birds may put more or less into reproduction depending on the quality of their partner.

A few years after this, one of my PhD students, Emma Cunningham, found that female mallards adjusted the size of eggs they laid depending on whether they were paired with a male that they had chosen themselves (i.e. a 'preferred' male) or one Emma had chosen for them (an 'unpreferred' male). However, she did not measure yolk size, so it is not known whether the quality of partner affects the amount of yolk that goes into an egg, but subsequent studies suggest that it does.[28]

You might imagine that a good way to test the idea that the amount of yolk affects chick quality and survival would be to experimentally alter the amount of yolk in an egg. But, as we saw in Chapter 6, this is more complicated and less informative than we might have hoped. The only way therefore is to look at the natural variation that exists in different bird species in the relative amount of yolk in their eggs.

Long before the idea of 'differential allocation' in zebra finches, or any other species, had even been thought of, Aristotle noticed that the amount of yolk in the eggs varied between different species of birds: 'The yolk is exactly circular and it varies in size according to the different size of birds, for waterfowl have a larger yolk and land birds a larger white.'[29] In 1884 the Russian physician and physiologist Duke Iwan Romanowitsch Tarchanoff, after examining the eggs of just nine bird species, recognised that the relative amount of yolk is closely linked to the state of development of the chick at hatching. Most small birds like blackbirds and robins whose chicks are blind, naked and helpless at hatching, produce eggs with relatively small yolks. In contrast, species whose chicks are precocial and can run around and feed themselves at hatching, such as chickens and ducks, have larger yolks. The extremes range from the northern gannet, whose yolk constitutes just 15 per cent of the egg, to the highly precocial southern brown kiwi where it amounts to an enormous 70 per cent.[30]

While it is intuitively obvious that the amount of yolk in an egg may affect the size or quality of the chick subsequently produced, it is less obvious that the *quality* of yolk might vary, and certainly less obvious that 'quality' may include hormones and other substances added by the mother.

The relative amount (percentage) of yolk in the eggs of different birds whose chicks hatch at different stages of development. From top to bottom: reed warbler (20%), gull (30%), duck (40%), mallee fowl (50%) and kiwi (70%). Redrawn from Sotherland & Rahn, 1987.

In the 1930s and 1940s, researchers found that, after inject-
ing laying hens with particular hormones, those same hormones
were then detected in the chicks hatching from their eggs. At the
time this was assumed to be a 'pathological perturbation', that is, a
non-adaptive effect. Then, in the early 1990s, a German biologist,
Hubert Schwabl, working at Rockefeller University in New York,
decided to see whether the fresh eggs laid by captive canaries and
zebra finches – which, crucially, had *not* been subject to any kind
of hormone treatment – contained hormones. Schwabl was interested in the chicks hatching from these eggs, so what he did was to
take a tiny sample of yolk using a syringe, and then seal the egg up
again and allow it to hatch. He found that the yolks of both species
contained testosterone, although the level in the canary eggs was
higher than that in the zebra finch eggs. Remarkably, the amount
of testosterone in the yolk was also completely independent of the
sex of the chick that eventually hatched from the egg. The fact that
Schwabl found the hormone testosterone in completely fresh eggs,
several days before there was an embryo large enough to produce its
own hormones, was convincing evidence that the mother had put
the hormone there.[31]

A second major finding was that in the canary each successively laid egg contained more testosterone in its yolk. Wearing
his evolutionary hat, Schwabl wondered whether the increasing
level of testosterone in successive eggs was an adaptive response
by the female, or the non-adaptive by-product of some other
process. A clue that the pattern might be adaptive was revealed by
the fact that the chicks hatching from eggs with the highest levels
of testosterone begged for food more vigorously than those from
low-testosterone eggs. It wasn't difficult to think of an adaptive
scenario: eggs laid later in the clutch sequence would hatch slightly
later and, because those chicks would be smaller, without some
intervention they would be at a disadvantage. The mother could

counter the hatching order effect by putting more testosterone into later laid eggs to create more aggressive chicks that might then be able to compete on equal (or more equal) terms with their earlier-hatched nest mates.

Demonstrating that testosterone levels in eggs is adaptive is tricky and Schwabl, and others inspired by his exciting research, knew that the only way to do this was to start wearing their physiologist's hats and begin to understand the mechanisms – the physiological processes – by which hormones, like testosterone, get into eggs. This approach raised many more interesting questions, including the concern that if the mother was putting hormones into eggs, how did those maternal hormones interact with those that the embryo eventually starts producing itself? Another question: if a female puts hormones into her eggs, does that mean she has to increase her own level of hormones to be able to do this, and, if so, does that create problems for her? One can imagine that if as a mother you wanted (in evolutionary terms) to put a high dose of testosterone into your eggs, it would entail making yourself more aggressive to do so. The answer to that particular question really depends on the mechanism by which hormones are placed in eggs. If it involves the female raising her own hormone levels, then it might be risky, but if the process were confined to the cells surrounding the developing ovum in the ovary, then the female herself would be unaffected. And that is precisely what was subsequently found.[32]

Coming back to the zebra finch, in the late 1990s Diego Gil and his colleagues at St Andrews University in Scotland paired females to males rendered either attractive with red rings, or unattractive with green rings, and looked at the amount of testosterone in their eggs. They found that females paired to attractive males had higher levels of the hormone in their eggs – providing clear evidence that female zebra finches adjust the content of their yolks according to the quality of their male partner.[33]

Despite this remarkable result, and twenty years after Schwabl's initial discovery, our understanding of the causes and consequences of hormones in yolk is still at an early stage. The patterns differ markedly between species, between different individuals and, as we've seen, between different eggs within a clutch, and even between different daily layers of individual yolks.[34]

The other type of substance female birds add to their yolks is carotenoids. It is these that make yolks yellow, and indeed the word 'yolk' comes from the old English 'geoloca' meaning yellow. Without carotenoids in their diet, the yolks of chicken eggs are white and it has always seemed a dreadful con to me that the poultry industry can legally add carotenoids to the food they feed battery hens to make their eggs more attractive 'to the housewife'. Carotenoids are essential for all sorts of biological processes, including embryo development, and in adult birds they are responsible for the red and yellow pigments in feathers, the beak and skin. Carotenoids, along with vitamins A and E, are antioxidants, and they minimise the damage caused by metabolism – so-called oxidative stress – to fats, proteins and DNA. Since birds cannot produce their own carotenoids they must acquire them through their diet. Because carotenoids are often scarce in the environment researchers have recently started to explore the idea that the amount of carotenoids and other antioxidants in egg yolks reflects a female's quality – or at least her ability to find them. It has also been proposed that the amount of antioxidants females deposit in particular eggs may, like hormones, reflect a way that mothers favour particular offspring. There's not actually much evidence for either of these interesting ideas at present, largely because there is little consensus across different studies, but one pattern does seem to be clear. The embryos of all birds grow relatively quickly, but different species grow at different rates. The faster an embryo grows the greater the potential for oxidative stress and the more need there will be for antioxidants. Charles Deeming and Tom Pike at the University of Lincoln in the UK found that species with rapid embryo growth had higher levels of antioxidants (carotenoids and vitamins A and E) in their yolks than those with lower growth rates. The sooty tern and Eurasian coot, for example, both produce similarly sized eggs weighing around 36.5g. The tern embryo grows at 0.89g per day, the coot more rapidly at 1.11g per day; the tern egg contains 280μg* of carotenoids in its yolk, but the coot egg contains 1180μg – over three times as much.[35]

There's still a great deal to discover about the role of both hormones and carotenoids in yolk.

Once the yolk has grown to its full size with its complement of whatever supplements the female may have added, it is ready to be released from the ovary, and to be fertilised. The vast bulk of the fully enlarged ovum – and remember this is a single cell – is yolk, but lying on top of that is the germinal disc, the tiny pale spot that comprises a small amount of cytoplasm and the female DNA – the stationary target of the male's sperm.

Most of us think of fertilisation as a particular and special instant: the moment a sperm penetrates the ovum. But deciding whether or when fertilisation has occurred is not quite that straightforward. The simplest definition of what constitutes fertilisation is rather like being at home, hearing a knock on the door, going to open it and seeing someone – a stranger, perhaps, or a close friend – and allowing them across the threshold. Another definition involves a scenario in which it is your lover at the door and while merely letting them in doesn't count as fertilisation, touching or embracing them does. In yet another definition, fertilisation involves both

* One microgram (1μg) is one millionth of a gram.

of you going upstairs, entering another room, getting into bed and making love – fusion, it's called.

As we'll see, the succession of events that occur between the sperm first entering the egg and what then happens to initiate the start of a new life is both sophisticated and wonderful. And they are very different in birds and people, even though the end result – the union of male and female sex cells – is the same.

In humans, the process of conception is the equivalent of going to the door and finding a single visitor, very occasionally two, only one of whom is allowed in. In birds fertilisation is more like opening the door and finding a football crowd, hundreds or thousands of people on the doorstep, and then having to decide who to let in. And it isn't a case of simply allowing a single select guest to enter; it is more a case of several guests gate-crashing, some through the door you've opened, but others getting in through the windows, only one of which eventually gets to embrace you as the host.

A source of fascination ever since the time of Aristotle, human fertilisation was always going to be a tough nut to crack, because everything happens out of sight deep inside the scientifically impenetrable oviduct. Even using other mammals – which researchers had little compunction about dissecting, as substitutes for women – working out how fertilisation occurs proved to be unbelievably difficult. The bird's egg seemed so much more tractable. The egg, inside which was the ovum, was something you could see and dissect, and in any quantity you wanted. And by keeping hens with or without a cockerel, you could control whether those eggs were fertilised or not. Yet, despite this fabulous accessibility, the process of avian fertilisation still remained frustratingly elusive.

It was William Harvey who made the first serious attempt to understand fertilisation. Having solved the great mystery of where the blood goes, he thought he could also discover the secret of where semen goes. He started with chickens. The union between a cockerel and a hen results in fertile eggs: all Harvey had to do was work out how. After allowing cockerels to inseminate laying hens, Harvey dissected the hens to find the semen, but he couldn't. It seemed to have evaporated. In frustration, and after many such attempts, he ended up imagining that fertilisation occurred rather like contagion – no obvious contact, but an obvious result. In his heart he knew it couldn't be true, but it was the only explanation consistent with the results of his experiments. As the king's physician, Harvey was allowed to dissect the female deer killed by royal hunting parties during the rutting season, but again there was no sign of the semen inside the hind's oviduct. It was perplexing, but at least it was consistent with what he'd seen (or not seen) in chickens.

It wasn't even like *Where's Wally?* (or, in the USA, Waldo), in which persistence eventually pays off. The problem was that Harvey was unaware of the microscopic nature of spermatozoa fervidly swimming inside the semen. There was nothing to see and for the deer he also had no idea what an ovum looked like, so it really was like fumbling for a needle in a haystack.

Understanding fertilisation was an intellectual obstacle course. A significant barrier in chickens was the fact that, even after separating male and female, a hen would sometimes continue to lay fertile eggs for several weeks. Fabricius erroneously thought that hens could produce fertile eggs for a full year, but Harvey correctly measured it as thirty days; but it was still a puzzle. In a similar way, the red deer and fallow deer that Harvey hacked open in the royal parks confusingly didn't show any sign of an embryo until two months after the rut had started. In the hen's case, successive fertile eggs were the product of sperm storage. In the king's deer, Harvey expected to see the seed of the male combined with the menstrual blood of the female, to create an egg-like structure in the oviduct. But deer don't have a menstrual cycle; Harvey had no clue of what a deer's ovum looked like, and, because of its unusual shape, failed to recognise the very early embryo for what it was, dismissing it as 'purulent matter'.[36]

Given his inability to solve these mysteries, it is hardly surprising that Harvey delayed publishing his study of generation. His previous work on circulation, although eventually confirmed as correct, had caused a huge amount of controversy. It is easy to understand why Harvey, now in his sixties, was reluctant to create another biological firestorm. It is also not surprising that he concluded that semen played virtually no role in generation, and why he used the quote *ex ovo omnia* – literally, *everything from the egg* – as the frontispiece to his book.[37]

The puzzle of fertilisation was eventually resolved by studying a much simpler system: frogs. Like hens, frogs were both abundant and immensely tractable, but crucially the effect of semen on an egg was obvious because it occurred externally, outside the female's body, and hence was visible. It was also immediate, with no hidden delays. Semen placed on a frog's egg resulted in an embryo within hours, as shown by George Newport in 1853. The actual process in which the ovum is penetrated by a single spermatozoa was finally revealed with an even simpler system, the sea urchin, by Oskar Hertwig in 1876, and, of course, by watching those events under a microscope.[38]

As with frogs, sea urchin fertilisation occurs externally following the release of both sperm and eggs into the sea and in response to an environmental trigger, such as a full moon or a particular sea temperature. Mammals and birds, in which fertilisation is internal, have evolved other strategies to ensure that male and female gametes come together. A particularly pragmatic solution is something called induced ovulation, in which copulation itself triggers ovulation. Cats, camels and rabbits are all induced ovulators. Another strategy is for females to signal, in no uncertain terms – usually through

A female grey wagtail (*right*) soliciting copulation from her partner.

scent or behaviour – their receptivity, which coincides with ovulation and is referred to as estrus, or being 'on heat'.

Humans are different. There's no estrus and ovulation is concealed rather than advertised. This means that women are in a state of more or less permanent sexual receptivity with copulation and insemination occurring pretty much all the time (well, relatively speaking), such that some copulations coincide with ovulation. The issue of what is referred to as 'concealed ovulation' in women has puzzled behavioural ecologists and reproductive physiologists for a long time.

Even though there are changes in a woman's appearance and behaviour around the time of ovulation, tests found that fewer than one-third of women know when they are ovulating, and most men are clueless. The most reasonable explanation for concealed ovulation in humans is linked to the idea that female reproductive success is dependent on paternal care. If women advertised their receptivity more males would be interested in mating with them and this would reduce their main partner's confidence in his paternity. He in turn would reduce the amount of care he provided for

her offspring. In addition, if her partner knew when she was ovulating and fertile, as soon she was no longer fertile he would go off and
look for other mating opportunities. If he then provided paternal
care to all the females he copulated with, each female would receive
a smaller amount of male care than if he had remained monogamous. In other words because paternal care substantially increases
female reproductive success, it pays women in evolutionary terms
to conceal when they are fertile because this enables them to secure
male help with child rearing.[39]

There are no known cases of copulation-induced ovulation in
birds; instead, females signal their fertile status by their willingness
to copulate. They do this by adopting a specific, soliciting posture,
often in response to male courtship or song. Copulation usually
begins a few days – in some cases several weeks – before the first egg
is laid. Here's what happens next. In response either to an internal
clock or to the presence of a male, the follicle that holds the ovum
ruptures, liberating it into the top of the oviduct. The portion of the
oviduct that receives, or in reality captures, the ovum is called the
infundibulum – which in Latin simply means 'funnel'. Although
'funnel' is a reasonable description, to me it conjures up an image of
a rigid, plastic kitchen funnel and the infundibulum is not like that
at all. Rather, this uppermost region of the oviduct more closely
resembles a diaphanous snake with an enormous mouth. If you see
the infundibulum of a dissected bird, it is hard to imagine how the
funnel opens to engulf the ovum, but it does so by creeping over
it, just as a snake consumes enormous prey by flowing over it. And
lying in wait inside the mouth of the funnel are hundreds or thousands of sperm.

It's all over in fifteen minutes. Regardless of whether fertilisation has occurred, after a quarter of an hour the ovum receives a
prophylactic covering – from the inner surface of the infundibulum
itself – effectively closing the door (or window) to any further sperm.

The bird's strategy is to ensure that the infundibulum is already well stocked with sperm, ready for that brief opportunity. The way birds do this is by copulating in advance of ovulation and storing sperm, thereby ensuring a ready supply for the moment when the ovum needs them. The sperm are stored further down the oviduct closer to the cloaca, in tiny, blind-ending tubes. A few days before the first ovum is released from the ovary, sperm start to be released from the tubes to be carried by the cells lining the oviduct up to the infundibulum, where they wait.[40]

As we've seen, chickens can store sperm for a maximum of thirty days. They naturally lay a clutch of around ten eggs (one each day), so thirty days provides a good safety margin. Pigeons lay just two eggs (albeit forty-eight hours apart) and store sperm for a maximum of just six days. But the number of eggs that need to be fertilised is only one of several factors affecting the duration of sperm storage. After all, some birds continue to copulate throughout the days that eggs are laid so storage is barely necessary. At the other extreme, prolonged sperm storage is absolutely essential for those species in which partners don't see much of each other.

Seabirds like albatrosses are socially monogamous and typically retain the same partner for their entire twenty- or thirty-year breeding life, but it's a curious relationship. The best way of thinking about it is to imagine that albatrosses are like long-distance truck drivers – at home with their partner only occasionally and making the most of it when they are.

Pioneering studies conducted in the 1990s, in which birds were fitted with transmitters and tracked by satellite, showed that albatrosses often seek food as far as 1,800km from their nest site. With food so far away, and with one partner having to remain at home to defend the nest site, it is inevitable that albatross partners cannot spend much time together. Male black-browed albatrosses return to their nest site at the start of the southern spring in late September;

the females come back about a week later and the couple are together at the colony for just one day.[41] A special day to be sure, because this is when virtually all copulations occur. The next day the female is gone, hundreds of kilometres away, gliding over the southern oceans in search of food that she finds mostly by smell, building up the reserves needed to produce her single egg. She remains at sea for just over two weeks, returning to the colony and her partner. With no further copulations, she lays her egg two days later.

In the past seabird biologists had a somewhat romantic notion of this period of absence from the colony, referring to it as the 'honeymoon period'. But this was a misnomer if ever there was one. For a start, it often involved the female going off on her own, and, second, it certainly didn't involve any of those activities that one normally associates with honeymoons. Better referred to as the pre-laying exodus, this is essentially a fishing trip – an extended one at that. In another albatross-like bird, the grey-faced petrel, the female is away from the colony building up reserves – and storing sperm – for a remarkable two months![42]

Albatrosses and petrels are extreme in how long they store sperm – in fact, they're extreme in many different ways – but most seabirds face similar problems precisely because their food is so far from where they breed. Guillemots have the same issue and I have spent several years working out what happens inside a female guillemot in the weeks prior to egg laying. Moreover, I have done this without ever once looking inside a female; instead, I have co-opted techniques developed by poultry biologists to establish the sequence of events.

In the weeks before egg laying, male guillemots spend much more time at the colony than the females. Females make brief visits, mainly to copulate. After this has happened a few times, they go off to sea to find the extra food they need to produce their egg. The female has already started to put yolk into one of her ova within

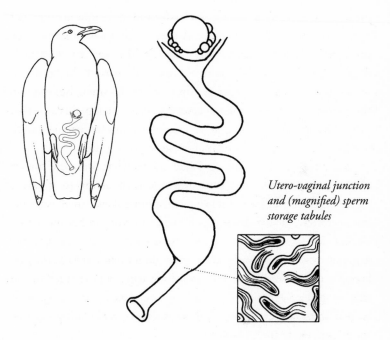

Utero-vaginal junction and (magnified) sperm storage tabules

A bird's ovary and oviduct showing the location (at the junction of the vagina and uterus) of the microscopic sperm storage tubules (*inset*) full of sperm.

her ovary, and two or three days before that ripe ovum is released from the ovary she releases the sperm from her sperm storage tubes. Like a vast crowd trying to get on the London Underground at rush hour, they surge en masse through her oviduct until they reach the infundibulum – equivalent to the Underground platform – where they wait for the train.

One of the most remarkable differences between the ova of birds and those of humans and other mammals is that mammalian ova require only a single sperm to start a new life, whereas birds, it seems, require several. A need for multiple sperm inside the ovum also seems to be true in a few other species including sharks, newts and salamanders, and is referred to as polyspermy, meaning literally many sperm. It is equivalent to the train arriving at the platform, opening its doors and allowing several of the waiting passengers to enter. Had this been a mammalian train, there'd have been room for just one.

Polyspermy *can* also occur in humans if there are too many sperm in the vicinity of the egg at the time of fertilisation, but with disastrous results. If too many sperm succeed in getting inside a human egg the embryo simply doesn't develop and for this reason it is referred to as pathological polyspermy. Normally, of course, the human oviduct – and this is true of other mammals, too – carefully regulates the number of sperm reaching the ovum, reducing the millions that were introduced at insemination to just one in the vicinity of the ovum.

On the rare occasions when there are several sperm in the vicinity of the human ovum, the ovum itself has a way of restricting access to only one. It is a process appropriately, if clumsily, referred to as the 'block to polyspermy'. Once one sperm has entered the human egg, a chemical reaction occurs that is rather like the Underground stationmaster giving the order to close the train doors and refusing to open them again regardless of how many passengers want to climb aboard. Sometimes, however, and most obviously during in vitro fertilisation – in a dish – there are a lot of sperm around the egg. In this artificial situation the sheer numbers of sperm overwhelm the egg and its block to polyspermy, breaking into the egg and effectively finishing it off.[43]

This isn't the case in birds, where, as we've seen, it is entirely normal for the ovum to be surrounded by hundreds or even thousands of sperm. Not only that, as was discovered in the early years of the twentieth century by American biologist Eugene Harper, several sperm routinely penetrate the ovum in the region of the germinal disc where the female genetic material resides. Because this does not result in the death of the embryo, it is referred to as physiological polyspermy.[44]

I find it remarkable that Harper's discovery of physiological polyspermy (which I'll refer to now simply as polyspermy) in birds

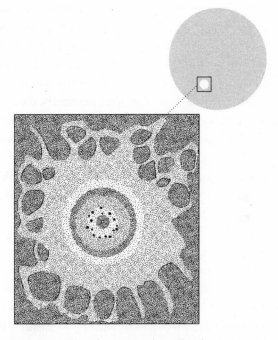

The germinal disc is the pale spot on the yolk. The main image is of the germinal disc (magnified) of a fertilised egg with the female pronucleus (containing the female's genetic material) located as the open circe at the very centre, and the surrounding dark spots are the several (16 in this case) supernumerary sperm, essential (in birds) for the successful development of the embryo after a single sperm has fused with the female pronucleus. Redrawn from Romanoff and Romanoff, 1949.

attracted so little attention. It is strange, too, that no one queried why mammals need only one sperm, but birds, along with sharks and some amphibians, seem to need several.

My colleague Nicola Hemmings discovered the answer while we were trying to devise a way of conducting artificial insemination in zebra finches. We wanted to inseminate females with a mixture of sperm from two different males whose sperm differed in length, because we were interested in knowing whether a 50:50 mixture of short and long sperm inseminated into a female resulted in a 50:50 ratio of sperm trapped on the tissue layer around the ovum. The way sperm length affects fertilisation success is another story;[45] the pertinent point here is that our efforts at artificial insemination failed and almost no sperm reached the infundibulum. It was frustrating because artificial insemination is routine in chickens and turkeys, and has apparently been successful in other birds as small as a zebra finch.[46] We were forced to find another way to test our sperm-length idea. But Nicola told me something that made me think. She said that even though most eggs had no sperm and were very obviously infertile, those with just one or two sperm had been fertilised but the embryo failed to develop.

At this stage we didn't know whether if we had left the eggs they would have developed, so we repeated the study and incubated them. None of the fertilised eggs developed: in every case, the embryo died just a few hours after fertilisation.

What this indicated was that in birds, polyspermy – the presence of additional sperm in the germinal disc – is a vital part of the process of 'generation', the term William Harvey and others used to encompass both fertilisation and development. Although the precise role of the supernumerary sperm in birds is not known, in frogs and newts it is thought that these extra sperm release substances that kick-start the embryo's development. Why this is necessary is not clear, but it seems to be true in birds, too.[47]

Minutes after fertilisation has occurred the ovum starts its twenty-four-hour journey down the bird's oviduct, where, as we have seen, the miraculous process of egg formation is completed. We have travelled with it, watching as each successive process has contributed to the creation of the most perfect thing. Here we are now, standing in the dark inside the uterus where the finished, hard-shelled egg waits like an actor in the wings, ready to make its appearance on life's stage.

8

Stupendious Love: Laying, Incubation and Hatching

A small crack in the shell is sufficient to destroy the surprising strength of the intact egg, the chick casts off the dried, now no longer useful, remains of the allantois and amnion, and steps out into the world.

Alfred Newton (1896)

In *Gulliver's Travels* Jonathan Swift describes a conflict between different factions within the kingdom of Lilliput over which end a boiled egg should be broken. By tradition the Lilliputians had always broken their eggs at the large end, but after the Emperor cut himself while opening the big end, he decreed that the little end should be the end for opening. This was not universally accepted and the quarrels over which end was opened gave rise to no fewer than six rebellions. Swift's endian wrangle satirises the ongoing eighteenth-century conflict between the Catholics (big-endians) and the Protestants (little-endians) over whether the body of Christ is actually or only symbolically present in the Host at communion.[1]

There has been a similar endian dispute over which way an egg emerges from the bird's cloaca: big end or little end first? Although there are some dissenters, most people – thanks largely to Aristotle – think that the blunt end emerges first. In turn this has led to an erroneous explanation for how the egg is propelled along the oviduct. Several early authors, including Friedrich Christian Günther, who wrote one of the first books on birds' eggs in the 1770s, assumed that because the blunt end emerged first that is how the egg travels down the oviduct. He suggested that it is pushed along by peristaltic forces, much like a food bolus in the gut, with the oviduct's circular muscles contracting behind the egg while those at the front are relaxed. It was thought that the trailing, pointed end – that is the longer portion of the egg – gave the oviduct wall greater purchase to squeeze the egg on its way.

One of the proponents of this idea, the nineteenth-century anatomist Heinrich Meckel von Hemsbach, was so confident about this explanation he called it a 'mathematical necessity'. To be fair, the idea does have an intuitive appeal, which probably explains why it was picked up and perpetuated by the great Scottish biologist D'Arcy Wentworth Thompson in his book *On Growth and Form* (1917). Thompson's intellectual stature was such that others assumed he must be right, including J. Arthur Thomson who repeated the error in his *Biology of Birds* published in 1923.[2]

I am intrigued by the way certain ideas in biology can persist for so long in the face of contradictory evidence. How could D'Arcy Thompson, J. Arthur Thomson and others ignore the evidence that flew in the face of their egg movement idea? As early as the 1820s two monumental figures in biology, the Czech biologist Jan Purkinje and the German Karl Ernst von Baer, both reported that even though the hen's egg usually emerges blunt end first, it passes down the oviduct *pointed* end first. Others confirmed that this was

Which end first? An egg as it travels down the uterus pointed end first
(*left*), then shortly before laying, turning (*middle*), so that it is laid blunt
end first (*right*). Redrawn from X-rays showing the rear end of a bird
with the outline of its body, part of the skeleton (pelvic girdle, the end of
the spine and one leg) and the egg itself. From Bradfield, 1951.

also true for pigeons, hawks and canaries, so why did Thompson and Thomson persist in their contrary view? Did they not believe their illustrious predecessors? Perhaps they didn't read German (for which they can be forgiven). The one paper that surely should have convinced them was by Heinrich Wickmann.[3]

Using eight very tame chickens that would lay their eggs on his desk, Wickmann recorded the events in the hours immediately before and during egg laying. Ingeniously, he was able to use a pencil to mark that bit of the egg he could see inside the hen's oviduct, through its cloaca prior to laying. (I can just imagine his wife popping into his study with a cup of coffee and seeing Wickmann with his pencil up a hen's bottom: 'What are you doing, dear?' she asks . . .). This allowed him to establish that, in the hour or so before it is laid, the egg is orientated with its pointed egg directed towards the bird's rear even though all eggs were all laid blunt end first. Wickmann deduced that the egg must turn immediately before it is laid.[4]

When I first heard of eggs turning in this way I imagined them doing so vertically, along their long axis – that is by 'pitching' – but they actually do so by rotating through 180° in the horizontal plane (i.e. yawing). This was discovered in the 1940s by John Bradfield who, you may recall from Chapter 3, used X-rays to observe hens' eggs on the later part of their journey through the oviduct. The broody hens sat immediately in front of the X-ray screen, and a succession of images was taken, starting at around midday just as the egg – covered only by the shell membrane – entered the shell gland. Images were taken, Bradfield says, at intervals until 9 p.m. and then restarted at 8 a.m. the next day. Had Bradfield been my PhD student I'd have suggested that he stay up all night at least once, although as it turned out it probably didn't matter. He wrote: 'That part of shell secretion which goes on during the night is unavoidably missed, but by following an egg which is ovulated early in the day it is possible to trace the first half of the process (which proved to be the most interesting), together with the last few hours.'[5]

When Bradfield examined his X-ray images what he saw was remarkable. An hour or so before laying, the shell gland with its fully formed shelled egg dropped a few centimetres out of the pelvic girdle, and over a period of just one or two minutes, during which the hen stood up, the egg rotated 180° horizontally. The dropping of the shell gland is possible because, unlike the mammalian pelvis which forms a circle of bone and through which the head of the foetus must pass at birth, a bird's pelvis is, as Bradfied says, shaped like an upturned boat, allowing the drop and rotation to occur, as well as facilitating the laying of large, hard-shelled eggs.

In each bird that Bradfield observed, the pattern was the same: the egg entered the shell gland pointed end first, turned and was laid blunt end first. Why turning should be necessary is unclear, especially for eggs like those of the domestic fowl and most passerines that don't actually differ all that much between ends. The fact that it isn't entirely consistent within species implies that the way the egg emerges cannot be that important. It is a pity that those researchers who observed the few chicken eggs laid pointed end first did not record whether those eggs were a different shape from those laid blunt end first. Perhaps for most eggs it is better – for some unknown reason – to undertake most of their journey down the oviduct pointed end first, but for the finale to emerge blunt end first is better.

According to Wickmann, after the egg has turned and is still lying within the uterus, something extraordinary happens: 'the egg is not simply forced out through the vagina as a baby mammal is born, the uterus is prolapsed through the vagina so that the egg is laid without having touched the walls of the vagina or the cloaca'. Based on his X-ray study Bradfield came to the same conclusion.

Soon afterwards, however, Allan Sykes, a poultry biologist who revisited this problem in the 1950s, was suspicious of Bradfield's conclusion: 'The only evidence we have for Wickmann's theory comes from Bradfield who claimed to have seen the uterus evert during a radiographic screening of oviposition [egg laying], although in his published radiographs . . . the oviduct is quite transparent to X-rays.' Sykes's own study provided no evidence for eversion of the uterus, and instead he found several birds with an egg lying in the vagina – which would be impossible if Wickmann or Bradfield were correct. Instead, Sykes suggests that the egg is pushed into the vagina much as a human foetus is in the second stage of labour by the contractions of the abdominal muscles. Once the egg reaches the opening of the cloaca, the hen bears down, much like a woman in labour, and then – with a bit of luck – the egg rapidly emerges.[6]

In birds, 'labour' may last a few hours in some species, or just a few seconds as it does in brood parasites like the common cuckoo (see below, page 185).

Guillemots, as is often the case, do it differently: their egg always emerges pointed end first and they are therefore 'little endians'. The female adopts a very upright posture with her legs in a vertical position (instead of the usual horizontal orientation); her neck is contracted into her shoulders and she holds her wings slightly away from her body. She can stand like this for ten or more minutes before the egg appears. When it does emerge, the pointed end often rests on the substrate while the rest of the egg remains in the female's body. She undergoes further abdominal contractions and then stands even more upright on tiptoe to allow the rest of the egg to fall out. As it does so, the female flops forward, catching the egg with her bill and drooping her wings to provide additional protection against the egg rolling away. After looking at the egg for a few moments – presumably to fix its appearance in her brain, or if she's bred before to remind herself what it looks like – she then positions the egg beneath her with the pointed end between her legs, and begins the lengthy process of incubation.[7]

We don't know whether the guillemot egg turns just before it is laid, as it does in other birds. Simply on the basis of the egg's size, this seems unlikely. If the guillemot is like other birds and the egg passes down the oviduct pointed end first, then given that it is laid in this way, there would be no point in it turning. What would clinch it? The answer is, being able to see inside a bird just before it laid. As I was writing this, I talked to my colleague Mike Harris who has studied guillemots for almost as long as I have, on the Isle of May, Scotland, and he told me how he once dissected a female guillemot that had just been killed by a predator: the egg was lying in the uterus with its pointed end directed towards the bird's bottom (cloaca).[8]

The guillemot isn't unique in the way its egg emerges; domestic ducks and geese, petrels and albatrosses, all of which produce much more rounded, symmetrically shaped eggs, also push the 'pointed' end out first.[9] But what about those other birds that lay distinctly pointed eggs, like waders and emperor and king penguins? To my surprise, the eggs of waders emerge pointed end first, but those of the two penguins emerge blunt end first, which does suggest that pointed end first is biologically significant in guillemots – as we'll see in the final chapter.[10]

Most small birds lay soon after sunrise in the morning. Possibly the first person to document this was the pioneer canary breeder Joseph-Charles Hervieux de Chanteloup in the early 1700s.[11] Tame birds in captivity provided an excellent opportunity to see when laying took place, although to what extent this knowledge filtered through to those interested in wild birds is unknown. Anyway, the first field ornithologist to note this early morning pattern of egg laying among songbirds was Alexander Wilson, a Scot who abandoned his homeland for North America in the 1790s and ended up producing the most comprehensive account of that country's birds, *American Ornithology*. For small birds, whose egg is relatively large, laying at sunrise seems like a good way of avoiding having to carry and risk breaking a fully formed egg in the uterus. A detailed study of the North American yellow warbler in the 1990s found that during the laying period females roost away from their nest, then come to the nest rather precisely, ten minutes after sunrise, and lay their egg within two minutes. It is possible that such a rigid schedule reflects the fact that the female has rather little control over when she lays.[12]

The common cuckoo provides a striking contrast, with females invariably laying during mid-afternoon. This was discovered in the 1920s by the amateur ornithologist and egg collector Edgar Chance, who was desperate to catch a female in the act of laying. At that time there was still a lot of controversy about how the cuckoo deposited its egg in its host's nest, with some people swearing that the female cuckoo laid her egg on the ground and then carried it in her beak to the host's nest. Others thought the cuckoo sat in the host's nest and laid its egg just as other birds do. It took Chance a while to figure out what was going on, but he finally saw and filmed cuckoos in the process of egg laying, confirming that they did what most other birds do, but rather more quickly.[13]

Subsequently, Nick Davies of the University of Cambridge provided an explanation for afternoon laying by showing that by depositing its egg at this time the cuckoo is less likely to be detected by the hosts, who are usually away feeding, having laid their own egg early in the morning, and are less likely to eject the foreign egg on returning to their nest. Because the cuckoo's egg is disproportionately small for the size of the bird, carrying the fully formed egg around for a full twenty-four hours, as it does, is unlikely to be much of a burden for the female cuckoo.[14]

Among larger birds, such as ducks, raptors and seabirds – including the guillemot – laying occurs at any time throughout the day. Seabirds like petrels and shearwaters that are nocturnal at their breeding colony, however, lay at night. For these birds where the egg is not particularly large relative to the female's body, carrying the egg and laying during the day or night may not be too much of an inconvenience and the egg may be less vulnerable to breakage. However, for guillemots, with an egg at 12 per cent of the female's body weight, the egg is quite bulky. It is also vulnerable to damage from the hurly-burly of a guillemot colony. Male guillemots have what one could call an 'impugnable appetite for copulation' and especially with the females of other males.[15] The benefits to males from a successful extra-pair fertilisation are obvious, the benefits to females far less so, and indeed they generally try to avoid such encounters, although it isn't always possible. In the period just before females start to lay, the colony is occupied mainly by males – randy males – and any female that arrives is likely to be pounced upon. If her partner is present, he helps to protect her, but sometimes she can struggle under a heap of several males. It is probably for this reason that females simply stay away from the colony in the few days just before laying, and return only as they are about to lay.[16]

When I mentioned to a friend that I was writing a book about eggs he told me to be sure to mention how in Thomas Hardy's *Jude the Obscure* the young Arabella incubates the sky-blue egg of a song thrush in the cleavage of her bosom. Although I had read *Jude* many years before, I didn't remember this, and my friend's comment conjured up false hope. When I checked, I was disappointed to find that it wasn't a song thrush egg, but that of a chicken, albeit a rare one called a Cochin, and instead of being brooded directly against the girl's skin, the egg was wrapped in a pig's bladder with some sheep's wool, 'in case of accidents'. The original image I had had in my mind disintegrated like the sound of a vinyl record after the power has been turned off.

Hardy's description of the bladder and wool suggests that, as with many of the quirky details in his books, it was based on something he had read in a newspaper. Assuming that someone really had tried to incubate a Cochin egg in this way, they may have been inspired by Pliny's account of Livia, the pregnant wife of Tiberius Claudius Nero, who, on asking a fortune-teller about the gender of her unborn child, was told that if she hatched a chick in her bosom its sex would be the same as that of her child.[17]

There were others seemingly inspired by Livia. René-Antoine Réaumur in his 1772 book *The Art of Hatching and Bringing up Domestic Fowls of all Kinds*, recounts how a woman incubated five goldfinch eggs in her bosom, four of which hatched after ten days – the fifth was addled and failed to hatch.[18]

There is also at least one case of a man hatching birds' eggs: the bird man of Alcatraz, Robert Stroud, whose life sentence for murder started in 1909, hatched house sparrow and canary eggs under his armpit inside a protective cup made from Elastoplast during his incarceration.[19]

The idea that humans could hatch birds' eggs was given further credibility by Lord Snowdon, the husband of Princess Margaret, in his 1969 television documentary *Love of a Kind* about people's devotion to pets, in which a chick hatches on camera from the cleavage of a sixty-year-old woman. But like so much on television it was contrived. The egg had not been incubated in situ; instead Snowdon had obtained a chicken egg on the point of hatching and placed it in position to film it.

One viewer wrote to say that what had been shown on Snowdon's programme was nonsense because a woman's body is not warm enough to successfully incubate a hen's egg. In response the television company consulted a poultry expert who confirmed that human body temperature was actually perfect for incubating birds' eggs. Even though the average body temperature of most birds is around 40°C, which is higher than the 37.5°C of people, because a bird's egg is incubated outside its body normal incubation temperature is around 36–38°C.

Réaumur's book was the outcome of an attempt to devise a way of artificially incubating hens' eggs. The ancient Egyptians and Chinese had both developed methods to warm and hatch eggs and hence maximise production: the French were keen to do the same in the eighteenth century and this was, in a way, the beginning of commercial poultry production. From the early 1900s there were numerous attempts to build effective artificial incubators. One, made in the 1850s by William James Cantelo, a farmer from Chiswick, became famous, partly because of its apparent effectiveness and partly because Cantelo advertised and exhibited it as a fairground event, with eggs going in at the top and chicks emerging at the bottom. It is not clear just how effective it was because the secret of successful artificial incubation is keeping the temperature constant and this was possible only with a thermostat – and thermostats were not used in this way until the 1870s. Cantelo may have been mainly a showman. As commercial and private poultry keeping expanded during the late 1800s and early 1900s, bigger and better commercial incubators continued to be developed: today, some incubators house over a million eggs. And, as with so many egg-related issues, a huge amount of information now exists on the best way to incubate chicken eggs. Of course, natural selection has ensured that wild birds do this properly; the poultry industry had to start almost from scratch in terms of temperature, humidity and turning.[20]

Temperature was the easiest to work out, and revealed, as we've just seen, that the optimum across most birds is 36–38°C. Under natural conditions and in most birds the heat comes from the brood patch (or patches) of the incubating parent – sometimes this is just the female, occasionally just the male, but in many cases both sexes. The brood patch is an area of skin from which the feathers are lost just before egg laying, and is well supplied with temperature sensors and blood vessels enabling the bird to regulate the temperature of the eggs.[21] The nest often plays an important role in incubation and nest construction has evolved in line with ambient temperatures, with better insulated nests in cooler regions.

One of the most beautiful and extraordinary examples of this is the goldcrest whose feather-filled nest of lichens and moss provides extraordinary insulation. It has to, for the goldcrest is tiny, just five grams – the same as a level teaspoonful of sugar – and extremely vulnerable to losing heat. The female alone incubates and she somehow has to warm a clutch of ten (relatively large) eggs even though her size allows her to be in bodily contact with only two or three at a time. In this case, a well-insulated nest isn't enough and it was Ellen Thaler, a German ethologist, who revealed the secret of the goldcrest's incubation strategy. Observing captive birds at close range, she noticed that when the female goldcrest took a break from incubation her legs were bright red rather than their normal brownish-yellow colour. When the bird returned to its nest ten minutes later its legs were brown again. Suspecting that the red colour indicated an increase in blood flow and a rise in temperature, she borrowed a device made by a colleague (to measure the body heat of flying locusts), and recorded the temperature of the incubating goldcrest's legs. To her amazement, inside the nest the legs registered at 41°C, but after ten minutes away from the nest this had fallen to 36°C. She realised that the incubating female pumps more blood through her legs to generate additional incubation heat among the eggs piled up beneath her. The legs and feet were effectively employed like an immersion heater in a water tank, keeping the eggs at a constant 39°C. Even more remarkably, as she incubates the female goldcrest gently paddles her legs to distribute the heat among the eggs, turning them at the same time. Thaler suspected that this must be what was happening and on placing a tiny microphone inside the nest she could hear the constant clunking of eggs as the female moved them around with her feet. A few other birds incubate with their feet, but the legs and feet of most birds are designed to *avoid* heat loss.[22]

At the other extreme, the guillemot incubates its egg on a bare rock ledge with no nest and no insulation. In Arctic regions – as I saw for myself – it isn't uncommon for guillemots to incubate with their egg resting on ice. Yes, they have a brood patch, but only a portion of the egg is in contact with that; the other side of the egg lies directly on ice or the rock ledge – not even on the bird's feet as in some other species. How they warm their eggs effectively under such conditions is unknown.

Among naturalists of the past the process of incubation incited complete wonder. Today we don't give it a second thought, but Willughby and Ray's comments from 1678 ought to give us pause for thought about just how remarkable the business of incubation is: that 'birds should with such diligence and patience sit upon their nests night and day for a long time . . . With what courage and magnanimity do even the most cowardly birds defend their eggs . . . Stupendious in truth is the love of birds to a dull and lifeless egg.'[23]

Having said that, despite the warning we are given as children that keeping an incubating bird off her nest will chill and kill her eggs, embryos are remarkably resilient. The eggs of most species can withstand a fair amount of chilling – which is essential if the incubating bird has to leave the eggs for any reason. The poultry industry maintains that fresh fertile chicken eggs can be left for about a week at room temperature without suffering any reduction in hatching success. We shouldn't be surprised at this because the domestic fowl's ancestor, the red jungle fowl, typically lays a clutch of about eight to ten eggs, one day apart, and doesn't start incubating until it has laid the last egg, so their eggs must remain viable for at least eight to ten days without incubation. However, as we saw in Chapter 6, the ability of recently laid eggs to remain viable depends on environmental conditions.

The conventional wisdom is that eggs with developing embryos are much more vulnerable to chilling than fresh, undeveloped eggs, but this varies between different groups of birds. The embryos of many marine birds that forage huge distances away from their nest seem to have evolved remarkable resistance to chilling. Fork-tailed storm petrels, breeding in Alaska, routinely leave their eggs unattended for days on end and one egg hatched after being abandoned for seven consecutive days. As a result, incubation periods in this species vary between thirty-seven and sixty-eight days – extraordinary flexibility.[24] Guillemots are assiduous incubators and leave their eggs only under extreme conditions, such as the approach of a predator or when food is desperately short. However, I once saw a guillemot egg, abandoned for four days, subsequently hatch after incubation was resumed.

What eggs withstand less readily are temperatures much higher than the bird's own body heat of 40°C. Eggs incubated at over 40°C rarely hatch. Indeed, this may be why birds incubate outside the body at lower than body temperature, rather than having embryos nurtured inside their body: at 40°C the rate of development might simply be too high to be healthy.[25] A few species of birds breed in extremely hot environments: the double-banded courser nests in the Kalahari and although it incubates its eggs at night when it is cool, during the day when temperatures are between 30° and 36°C it merely stands over them to shade them from the sun. When air temperatures exceed this the bird – paradoxically – resumes incubating again to *prevent* them getting too hot. The Egyptian plover – which is neither Egyptian nor a plover – nests on open sand bars on rivers fully exposed to the tropical sun. In the 1850s the German ornithologist Alfred Brehm made the remarkable discovery that this species buries its eggs in the sand, encouraging some ornithologists to speculate that it breeds like a megapode. Others wondered whether egg burying might be a way of keeping the eggs relatively cool, noting that the surface of the sand could reach over 45°C in the middle of the day. Consistent with this idea was the observation that the adults regularly walk to the water's edge, dip their bellies into the river and return to the nest to drip water over their buried eggs. Then, in the 1970s, detailed observations of Egyptian plovers by American ornithologist Thomas Howell allowed him to conclude: 'A balanced combination of body heat, solar heat, and heat retained by the sand keeps the TE [incubation temperature] within appropriate limits.' Burying (and wetting) helps to prevent the eggs from overheating, but also conceals them from predators (as it does with the chicks), and allows these birds to exploit an unusual habitat.[26]

In commercial operations, chicken eggs are artificially incubated at about 50 per cent humidity, which seems to optimise hatching success. The reason why the humidity needs to be controlled is that it affects the rate of water loss from the egg; too much or too little humidity can kill the embryo. Among wild birds, in contrast to temperature, which is actively regulated by the behaviour and physiology of the incubating bird, humidity is not. Instead, humidity is controlled passively through choice of nest site or habitat and the evolution of eggshells that function effectively in particular environments or, as we saw with the changes in eggshell structure, in response to altitude (Chapter 2), through a physiological flexibility that allows the female to produce an appropriately designed shell. The only birds that need to actively ventilate their nest sites, to ensure an appropriate level of humidity and oxygen, are those breeding in tunnels. Incubating woodpeckers and bee-eaters shuttle backwards and forwards along their tortuous nesting tunnels several times each night to push air in and out to ventilate the nest chamber.[27]

'Turn! Turn! Turn!' was a hit for the Byrds in 1965. Had they been able to understand, this would have been appropriate advice for any incubating birds, for turning is essential for successful hatching. Réaumur in his eighteenth-century incubation monograph sensibly suggests that turning ensures that all the eggs in a clutch are adequately warmed. But in the 1950s it was discovered that this wasn't true: poultry eggs in commercial incubators that generated an extremely uniform warming environment still failed to hatch unless they were turned. As long ago as the 1890s it had been suggested that without turning the embryo became stuck to the shell membrane, causing its death. This particular idea was subsequently widely accepted, and commercial incubators with facilities for regular turning were developed. When the physiological consequences of not turning eggs were explored in the 1980s and 1990s, Charles Deeming, now at the University of Lincoln, discovered that the idea of embryos becoming stuck to the shell membranes was wrong. Dead embryos stick to the shell membranes, but sticking was not the cause of death. Rather, embryos died because they were unable to properly utilise the egg's albumen. What turning does – and it is only absolutely essential for a few days in the early part of incubation – is to promote the development of the embryo's network of external blood vessels, encouraging the diffusion of nutrients and water within the egg, and ensuring that the embryo is optimally positioned with respect to the yolk and albumen, all of which allow it to make full use of the albumen as it develops.[28]

As the embryo completes its development its weight distribution within the egg becomes uneven so that lying in the nest the egg has a distinct lower (heavier) surface and an upper surface. The parent birds' 'turning' behaviour now serves *not* to turn the eggs but to retain their orientation so that when the chick first breaks through the shell it does so on the upper surface which is free from any vegetation or other obstruction in the nest.[29]

There are several birds that cannot turn their eggs. These include the palm swift, which glues its two eggs with saliva into its funny little nest, which is itself glued to the underside of a palm leaf. Megapodes, like the malleefowl and brushturkey we met previously, are also unable to turn their eggs because they are buried in soil or vegetation. Nor do kiwis turn their enormous egg, simply because there is insufficient space in their nesting burrow to do so. The palm swift may be easily explained, since the flapping of the palm leaf in the wind probably provides sufficient movement to compensate for any lack of turning. The megapodes and kiwis pose a different problem: how do they manage to hatch their eggs without their being turned? Charles Deeming provided the answer. By recognising that turning facilitates the embryo's use of albumen, he predicted that eggs with relatively large amounts of albumen – laid by birds that hatch altricial young (those hatching blind and naked) – would require more turning than those eggs giving rise to precocial young (those with their eyes open and capable of running around). Not only does this seem to be the case, it also explains why both megapodes and kiwis – and reptiles – which have relatively tiny amounts of albumen in their eggs, can get away without turning their eggs. The eggs of small songbirds whose chicks are altricial contain about 80 per cent albumen; for ducks whose chicks are precocial the albumen makes up about 60 per cent of the egg contents; for the super-precocial malleefowl it is 50 per cent and for the extraordinarily precocial kiwi it is just 30 per cent. Reptiles fall between these last two bird species with their eggs composed of about 45 per cent albumen.[30]

It has been known since before the time of Aristotle that a chicken's egg requires twenty-one days of incubation. For most other birds the duration of incubation remained pretty much a mystery until the twentieth century. Only in the early 1900s, when Oskar Heinroth and his wife Magdalena started to hatch eggs in incubators and foster them under canaries or geese in their Berlin apartment, were accurate measurements obtained. Over a twenty-eight-year period this remarkable couple reared over a thousand individual birds of three hundred different species. In the process, Oskar, who was assistant director of the Berlin Zoo, made numerous novel discoveries about bird behaviour that subsequently have been largely credited to others. Tragically, Magdalena died just two weeks after the end of their huge project, and before its results could be published. When their monumental four-volume account *Die Vögel Mitteleuropas* [The Birds of Central Europe] finally saw the light of day, it was overshadowed by the onset of the Second World War.

The shortest known incubation period – and this is the time it takes to go from laying a fertilised egg to a hatched chick – is just ten days in some small songbirds. The longest incubation period is around eighty days in the royal albatross and kiwis. Very broadly, the larger the egg, the longer it takes to hatch. There's a lot of noise, so to speak, in this relationship and species with similar sized eggs can often differ considerably. As the Heinroths noted, the egg of a griffon vulture weighs 250g and takes forty-nine days to hatch its altricial young, whereas an ostrich egg weighing 1,500g produces a precocial little ostrich after just thirty-nine days.[31]

The incubation periods of birds are determined by a combination of their evolutionary history and their ecology. For example, all members of the bird family known as procellariformes – shear-waters, petrels and albatrosses – have relatively long incubation periods: that's an evolutionary history effect. As seabirds, their chicks (including the time they are embryos while inside the egg) grow only slowly, because food is a long way from the colony: that's an ecological effect. Birds that breed in cavities – such as tits and chickadees – where they are relatively safe from predators, also tend to have long incubation periods: an ecological effect. Overall, as the Heinroths recognised in the 1920s, long incubation periods occur in those species where the nestling period is also long, indicating that the rates of development both in the egg and after hatching are under similar genetic control.[32]

Hatching is the climax of incubation; indeed, it is the climax of both fertilisation and incubation and the third great landmark in the life of an egg. How does the chick break out from the claustrophobic confines of the shell? Our mental image of the process has been corrupted by cartoons, where attempts to romanticise and sanitise the process often show a hen's egg with its top neatly popping off to reveal a warm, yellow fluffy chick. The reality is not like that. It is still pretty remarkable, but it isn't as quick, as clean, or as simple as we have sometimes been led to believe.

A fully developed embryo lies scrunched up inside the egg with its ankles at the pointed end and its head towards the blunt end; its neck is bent so that the head lies adjacent to the breast with the beak poking out from under the right wing up against the egg membrane. This pre-hatching posture seems to be typical of all birds, except for megapodes.

Before starting to break out of the egg the chick has three things it must accomplish. It must first switch from being dependent on the oxygen diffusing through the pores in the eggshell into the network of blood vessels that line the inner surface of the shell and start to use its own lungs to breathe. The chick takes its first proper breath and fills its lungs the moment it punctures the air cell inside the top of the egg. This step is essential because by this stage of development there is not enough oxygen diffusing through the pores in the shell to support the chick's respiratory requirements. Taking a breath from the air cell provides the oxygen and the energy necessary to break through the eggshell.

Before it takes that first breath, the chick has to start shutting off the blood supply to the network of blood vessels that line the inner surface of the shell, and withdraw that blood into its body. The blood vessels are programmed to close off at the point where they emerge from the bird's umbilicus, and just before the chick starts cutting round the shell.

Third, the chick has to take what is left of the yolk and draw it into its abdomen. It does this by sucking up the remaining yolk through the stalk that connects the yolk to the chick's small intestines. This 'yolk sac' is a food reserve for the first few hours or days after hatching.

Essentially, the chick has to do what a human baby does as it switches from dependence on the placenta for both oxygen and food to independent breathing with its lungs and the ingestion of food through its mouth. Thinking of it like that, it is a pretty major transition.[33]

The chick is now ready to break out of the shell and starts by thrusting its beak against the inside wall of the shell. To help puncture it the chick employs a tiny structure of especially hardened material at the tip of the bill. Known as the egg tooth, its role in hatching was discovered by the ornithologist William Yarrell in 1826. Watching domestic ducks and hens hatching in an early incubator, and by removing a fragment of shell, he could see the

sharp little egg tooth pressing against the inside of the egg, ultimately enabling the chick 'by its own efforts to break the walls of
its prison'.[34] Reptiles (including at least one dinosaur) also have an
egg tooth, as do the egg-laying mammals, the duck-billed platypus
and echidna: it is the key for getting out of a shell. In birds, the egg
tooth is made of calcium and is usually confined to the tip of the
upper mandible, although some species such as avocets, stilts and
woodcock have an egg tooth on the tip of the upper and lower beak.
In most birds the egg tooth falls off a few days after hatching, but
in passerine birds (such as finches and sparrows) it is absorbed back
into the bill. In petrels, the egg tooth remains visible for up to three
weeks after hatching.[35]

As it breaks through the shell the chick takes its first breath of
atmospheric air: its first breath of air outside the shell. Energised
by this pulse of extra oxygen, the chick continues to peck away at
the inside of the shell and simultaneously starts to press its shoulders and legs against the inside of the shell. It also begins to rotate
its body inside the shell in an anti-clockwise direction (if you are
looking down on the blunt end of the egg). The egg tooth then
makes a hole in the eggshell, a process known as 'pipping'. I suspect
it was originally called 'peeping' after the noise the chick makes at
this point, since there's a note in Fabricius's account of the development of the chick from 1621 entitled 'Peeping is a sign that the
chick wishes to leave the egg'. As pipping continues, it eventually
results in the top of the eggshell, above the widest point of the egg,
falling free and allowing the chick to emerge. This is the commonest way that chicks get out of eggs. In a few species the chick splits
the side of the egg and emerges through an untidy hole – a method
of hatching that seems to be most common in birds with longish
beaks, like waders.[36]

Megapodes are different. Incubated in warm soil or fermenting
vegetation, they can afford to have a relatively thin shell because

their eggs don't have to bear the weight of an incubating parent. Also, because each egg lies in glorious isolation in its incubator, there's no risk of their being damaged by colliding with others or being kicked or pecked by the parent bird. The megapode's thin shell facilitates gas exchange, but it also means that breaking out is relatively easy. Megapode chicks don't have an egg tooth, although one does appear – like an evolutionary ghost – early in development only to disappear by the time of hatching. Instead, megapode chicks hatch feet first, kicking their way out of the shell. To avoid injuring themselves as they hatch, the chicks' sharp claws are covered by jelly-like caps that fall off soon after they emerge above ground. A further difference is that megapodes start to breathe air as soon as they break through the eggshell because the business of digging themselves out of the soil or vegetation, which takes around two days, is energetically demanding and requires a good supply of oxygen. It was once thought, presumably because they were also buried, that dinosaur eggs hatched in a similar way to megapode chicks, but the discovery of an egg tooth on one of the extremely rare fossils of near-hatching dinosaur embryos suggests that this is not true.[37]

In a wide range of birds from owls to budgerigars, the parents sometimes help their chicks out of the egg by breaking off bits of shell at the point where it is penetrated by the chick's beak. In other species, the parents help by tipping the chick out of the shell once the cap is removed.

Among those birds that cut the top off the eggshell, some, like the ostrich, cut through no more than a quarter of the egg's circumference before shattering the shell and breaking out. At the other extreme, barn owls, pigeons and quail cut right round the top of the shell, neatly removing the entire cap before emerging. The bobwhite quail, which also removes a complete cap, even goes round more than once.[38]

Researchers have speculated about why there should be such variation in the way different bird species emerge from the shell. One idea is that hatching might be influenced by the degree of development, with precocial species, like chickens, being stronger and more able to break out of the shell than altricial species, like blackbirds and robins. But this idea seemed unlikely as species with precocial chicks include those that cut both the smallest (ostrich) and the largest (quail) pipping perforation tracks before emerging. Much more plausible is that eggs whose shell membrane and eggshell are tough and flexible require more cutting before the chick can escape than eggs that are hard and brittle. The eggs of ducks and chickens are hard and require only a few pips to destabilise the shell's integrity, and chicks can emerge after a relatively few pips. Quail, pigeons and the guillemot, on the other hand, have less brittle, relatively tough eggs and membranes and require more perforations to release the chick.[39]

The final, climax phase of hatching, in which the chick emerges from the shell, varies from a few minutes in small songbirds to a

Two extreme stages of chick development at hatching.
Left: a typical passerine (songbird) hatches, blind, almost naked and helpless (referred to as 'altricial'), and right: a plover chick hatches with its eyes open and is capable of running around and feeding itself (referred to as 'precocial').

day or more. In the chicken, the chick punctures the air cell about thirty hours before hatching, makes its first pip of the eggshell at twelve hours before hatching, and starts to rotate within the shell just fifty minutes before it emerges. In the guillemot, the air cell is punctured thirty-five hours before hatching; the first pip appears at twenty-two hours, and rotation starts about five hours before the chick emerges. As well as the effort required to cut through the relatively thick shell membrane (120µm) and shell (500µm), there is another reason for the more protracted process in the guillemot – the chick and its parents have got to be able to recognise each other's voices *before* the chick hatches. Remember that guillemots live beak by jowl with their neighbours at incredible densities and with no nest. They can recognise their own egg (Chapter 5), but they also need to be able to recognise their chick and it may take a couple of days to complete that process. Soon after the guillemot chick breaks through into the air cell, it starts to peep, and there is something magical about hearing a guillemot chick inside a still intact egg and its parents calling in response. Their individually distinct calls create a bond between the parents and the chick that ensures they can recognise each other the moment the chick breaks free from the shell. In the closely related razorbill, which breeds in solitary sites away from other razorbills, such immediate parent–offspring recognition does not occur because there's no risk that chicks from different families will become mixed up.[40]

I am thrilled by the idea of a guillemot chick inside its egg communicating with its parents. But in birds producing clutches of eggs that give rise to precocial chicks, something even more remarkable happens. In such species it is important that all the chicks hatch at the same time and can be taken en masse by the mother to safety. Female ducks, for example, minimise delays between the hatching of successive eggs by starting to incubate only once the entire clutch is complete. Nonetheless, some embryo development occurs even with no, or minimal, incubation, suggesting that the spread of hatching times might still be considerable.

One of the many novel observations made by Oskar and Magdalena Heinroth was that mallard ducklings from the same clutch hatched with extraordinary synchrony – over just a two-hour period. Despite this remarkable observation, no one thought much about synchronous hatching for a further forty years until another German ornithologist, Richard Faust, reported the same phenomenon in captive American rheas. Even though the interval between laying and hatching in different rhea clutches varied from twenty-seven to forty-one days, the chicks still hatched over just two or three hours. Faust realised that something must be causing this synchronisation but he did not know what.[41]

Margaret Vince, a researcher in Cambridge during the 1960s, solved the problem when she discovered that eggs talk to each other. She noticed that if she held a Japanese quail egg close to her ear just before it hatched, she could hear a peculiar clicking noise. This sound is uttered by the chick between ten and thirty hours after it has first pipped the shell and Vince realised that this might be how eggs in the same nest signal to each other and synchronise their activities. To test her theory she reared bobwhite quail under different circumstances and found that the eggs must be touching for synchronous hatching to occur, suggesting that the communication is partly auditory and partly tactile. Indeed, when she exposed quail eggs to artificial vibrations or clicks, she could induce synchronous hatching. The chick's clicks could either slow down or speed up the hatching process in adjacent eggs: most remarkable of all, when Vince added an egg to a clutch twenty-four hours later than the others, it was able to speed up its hatching to such an extent that the chick emerged from the egg at the same time as the others.[42]

The chicks of different bird species hatch in various states of development. At one extreme are helpless 'altricial' chicks of song-birds; at the other are the completely independent 'super-precocial' chicks of the megapodes which hatch fully feathered, their eyes open and capable of flight. In between, there is the familiar baby chicken – eyes open, covered in down and, although capable of feeding itself, still dependent upon its mother for protection and care. The guillemot chick is slightly less precocial than this, in that while its eyes are open and it is covered in down, it cannot run around and it cannot control its own body temperature. And prob-ably just as well: cliff ledges are no place to be running around, at least not until the chick has some decent coordination and a good sense of what an edge is – which it acquires as it grows. Because the guillemot chick is unable to maintain its body temperature, it requires warming against its parent's brood patch, which also helps to keep it safe.

What's left as the chick hatches? The answer is, not much: just the shell, which is slightly thinner than it was when the egg was laid because the chick has taken some of the calcium to form its skel-eton. But the empty shell is a liability: its sharp edges could injure the delicate young chick; the chicks could be trapped inside a shell; but worse, the pale-coloured inside of the shell makes an egg that was once cryptic highly conspicuous to predators. The parents deal with these challenges in one of two ways: either they eat the shell or they remove the eggshell from the nest. Most commonly the parents carry the two pieces of eggshell away. Birds like herons, nesting high up in trees, simply flick the shell pieces out of the nest; grebes, which nest on water, push the shell pieces out of sight beneath the surface; and ground-nesting birds like gulls pick the pieces up in their beak and fly off before dropping them a few tens of metres away.

In an elegant set of field experiments on nesting black-headed gulls conducted in the 1950s and 1960s Niko Tinbergen demonstrated both the cues that stimulate eggshell removal and the survival value of eggshell removal behaviour. The cue that triggers removal is the light weight of the empty eggshell; and the survival value is that it removes the white inner shell that predators like crows cue in on to find tasty young chicks. Ducks simply leave the eggshells in the nest but remove their synchronously hatched chicks to places where they are safer from predators. Guillemots and other cliff-nesting birds, like the kittiwake, simply leave the eggshell wherever it is, because their chicks are relatively safe from predators.[43]

In the next, and final, chapter, we shall draw together the several strands we have explored in this book.

9

Epilogue: Lupton's Legacy

For my own part, that taste for natural history which I have enjoyed almost from the earliest recollections, has proved to me an inestimable blessing. To its influence I owe all the brighter hours of my life.

William Hewitson, *British oology* (1831)

Short of money as he often was, George Lupton sold his extraordinary collection of guillemot eggs some time around 1940 to an idiosyncratic millionaire, Vivian Hewitt.

Born in 1888 into a massively successful Grimsby brewing empire, the young Hewitt, with a passion for fast cars and aviation, had three claims to fame. The first was that, on 26 April 1912, he flew from Holyhead in North Wales to Dublin, Ireland, on what was claimed to be the first and longest – seventy-three miles – crossing of the Irish Sea by air to date. His one-hour flight in a wood, wire and canvas glider was extremely risky for, in the previous year alone, no fewer than sixty-five would-be aviators had perished trying to master the newly invented flying machine. Hewitt was twenty-four and his successful endeavour made him a celebrity.[1]

Hewitt gave up flying soon after the First World War for health reasons. He inherited the brewing fortune and, with no need to earn a living, needed something to occupy him. He bought boats and began to visit the Welsh seabird islands, including Bardsey and Grassholm, to collect birds' eggs. He was probably kept clear of Skomer's cliffs by the owner, Reuben Codd, who was ferocious in his protection of the island's wildlife. Like Lupton and many others, Hewitt also bought guillemot eggs from the Bempton climmers, often paying ridiculous prices for unusual specimens. They must have loved him![2]

In the 1930s Hewitt purchased a property at Cemlyn, near the beautiful northern tip of Anglesey, where he developed a bird reserve. After a decade of collecting eggs for himself, Hewitt switched to buying entire egg collections from others, and hence became 'a cabinet man'. It was the accumulated collections of other peoples' eggs that was Hewitt's second claim to fame.

Despite his phenomenal wealth, Hewitt's domestic life was fairly basic, for Cemlyn had neither running water nor electricity. The contrast to the luxury he enjoyed at the Savoy Hotel in London, where Hewitt was a regular guest, could hardly have been greater. He never married – possibly because of a domineering mother – but instead was cared for throughout his entire adult life by two loyal retainers, Mrs Parry and her son Jack. The three of them lived reclusively behind the enormous wall Hewitt had built around the property, allegedly to protect the birds within its grounds.

As well as eggs Hewitt bought bird skins, engines, coins, stamps and guns, and Cemlyn became a huge storage facility for his vast, untidy collections. Financially unrestrained, he procured whatever he wanted and often on a whim. His most famous purchases were of several mounted great auks and great auk eggs. With the last known specimens killed a century earlier, the skins and eggs of this species were incredibly scarce – and expensive. Learning that there was a stuffed great auk and an egg in a Scottish castle, Hewitt sent another egg collector, Peter Adolph, armed with a blank cheque to get them both. Upon his arrival, the laird told Adolph that Hewitt wouldn't be able to afford them, but the price the laird stated was actually 75 per cent less than what Adolph had anticipated. So the deal was done. In total, Hewitt acquired four stuffed great auk specimens and no fewer than thirteen eggs: more than any other private collector before or since. That was his third claim to fame.

One of the few people to witness life inside Cemlyn was Hewitt's doctor, William Hywel, who, after Hewitt's death, was also his biographer. Hywel summed Hewitt up with the words: 'Once he inherited his wealth all previous ambition evaporated: so much started, so little completed.' It was obvious that acquisition was everything. He had no idea about curating his collection and many of the cases of eggs that he bought from others remained unopened. We don't know whether he ever looked at Lupton's collection, which he held for more than twenty-five years. If he did, it may have partly been Hewitt's carelessness that led to the sad chaos that I found in the egg rooms at Tring.

When Hewitt died in 1965 Jack Parry inherited the entire collection of half a million eggs. He didn't want them and was advised to push the whole lot over a nearby cliff. Luckily, he was persuaded to donate them to the British Trust of Ornithology (BTO).

It took four huge removal lorries to transport the eggs to the BTO's headquarters in Tring. With insufficient space, the BTO asked the Natural History Museum, also at Tring, if they would store them. The museum agreed, but only on a temporary basis, for they were about to start renovations and would soon need the space. Word of the collection got out and before long a certain

Johnnie du Pont, another millionaire oologist and founder of the Delaware Museum of Natural History in the USA, came to look. The director of the BTO at the time, Jim Flegg, remembers du Pont arriving in a Rolls-Royce accompanied by a chauffeur and two minders. On being shown the collection, du Pont offered to buy it and made an offer. Flegg, who was desperate for funds for the BTO, told him to double it – which he did: £25,000. It isn't clear whether du Pont knew that the BTO had already allowed the Natural History Museum and some provincial museums to take what they wanted from the collection. It also isn't clear whether du Pont knew that the best items – including the great auk eggs – were in the Bahamas, where Hewitt had a second home.[3]

Soon after the du Pont sale was agreed, British oologists, members of the Jourdain Society, recognised that Hewitt's collection contained eggs that once belonged to their founder, Francis Jourdain, and insisted that this collection remain in the UK. The result was that *some* of the Jourdain collection was retained by the Natural History Museum. But, as the curator at Tring told me: 'This badly organised collection split led directly to a mix-up of data and a proportion of eggs from the Jourdain Collection are in Delaware and a series of eggs in Hewitt's collection (i.e. some of Lupton's eggs) are in Tring. The resulting confusion has left a residue of eggs with no data slips, and data slips without accompanying eggs in both institutions. It is not easily sorted.'[4] I contacted Jean Woods, the curator at Delaware, and she confirmed that their collection of guillemot eggs is in as much need of some tender loving care as the one at Tring.[5]

George Lupton died in 1970, some fifteen years after suffering a stroke. His family sent me a faded colour photograph of him in old age. Sitting low in a battered leather armchair, he's wearing a collar, tie and a sports jacket. The walls of the room are covered with paintings and photographs, including an image I was familiar with: a coloured plate from Henry Seebohm's *Eggs of British Birds* (1896) showing six guillemot eggs.[6]

I was twenty when Lupton died and I'd only recently seen a guillemot, but I like to think that Lupton would be pleased by the fact that I have used his collection to help me understand several aspects of bird biology, including the shape of their eggs.

So why do guillemots produce such an unusually shaped egg? The fact that large numbers of guillemot eggs tumble off the breeding ledges when a predator frightens the birds suggests that their conical shape isn't much of an adaptation to prevent rolling. Paul Ingold's rolling experiments (Chapter 3) provide little support for the rolling-in-an-arc hypothesis. Neither does the fact that the radius of that arc – at 17cm in fresh eggs and 11cm in well-incubated eggs – is greater than many of the ledges on which guillemots breed. In addition, we found no convincing support for Ingold's idea that larger (heavier) eggs of both guillemot species are more pointed (Chapter 3). All of this isn't to say that the shape of a guillemot egg has no benefit with respect to rolling, but it does suggest that rolling is not the main factor in the evolution of guillemot egg shape.

Making a judgement about what is adaptive is not straightforward. All biologists know that no adaptation is perfect, and the reason for this is that what evolves is invariably a compromise between different selection pressures. For the common guillemot, the compromise is between two major selection pressures: rolling, which has been the focus of most researchers' work so far, and muck – faecal contamination – which until now has barely been considered. In my opinion, and based on our research, the most likely explanation for the guillemot's pointed egg is that this shape keeps the blunt end out of the muck. It may also be why it is laid pointed end first. When we looked at the distribution of faecal contamination on guillemots' eggs, most of it was on the pointed end, and in most – but not all – cases the blunt end was clean. The blunt end is where the embryo's head lies; it is where the air cell resides and it is where the diffusion of air through the shell is most important. It is the region from which the chick has to emerge.[7]

Who would have imagined that identifying the adaptive significance of guillemot egg shape could have been so difficult? Although my research provides a new perspective, I'm experienced enough to recognise that the story might still not be over.

The study of birds' eggs may seem an indulgence. Who cares about eggs? A lot of people care about birds' eggs being protected from collectors – a concern motivated by conservation. But the study of eggs can aid conservation, too. Overall, about 10 per cent of all eggs laid by wild birds fail to hatch. Given that each egg, with its huge yolk-filled ovum, represents a massive investment by the female bird, throwing 10 per cent away seems extraordinarily wasteful. Among endangered birds, the proportion of eggs that fail to hatch is even higher. In the kakapo – the giant, beautiful, flightless green parrot of New Zealand – for example, over two-thirds of all eggs fail to hatch. While hatching failure is not usually the cause of a species being endangered, it exacerbates their plight.

Ornithologists usually refer to eggs that fail to hatch as 'infertile', but that is misleading because an egg can fail to hatch for two main reasons. The egg may be genuinely infertile, that is unfertilised, either because of insufficient sperm or indeed a complete lack of sperm. But eggs may also fail to hatch despite being fertilised, because the embryo dies. This cause of hatching failure is easy to identify if the embryo dies after it is several days old; that is, several days after fertilisation. Such eggs are often referred to in the poultry industry or by bird-keepers as 'dead in shell'. It is when the embryo dies soon after fertilisation that confusion arises, because the contents of an unhatched egg in which there has been early embryo mortality are – to the human eye – identical to those of an unfertilised egg.

The difference is important, especially if you are trying to understand why the eggs of endangered birds fail to hatch. Unfertilised eggs are generally indicative of a male problem: unable to make sperm; unable to transfer sperm to the female; unable to produce sperm capable of completing the journey to the infundibulum; or sperm unable to fuse with the female gamete to form an embryo. If, on the other hand, hatching failure is due to early embryo mortality, the problem could lie with the female, or more probably with the genetic incompatibility between partners. In humans genetic incompatibility is known to be a cause of early embryo mortality and spontaneous abortion.[8]

Because the term 'infertile' has been used so indiscriminately to describe the unhatched eggs of birds, it is generally assumed that every instance of hatching failure where there is no obvious embryo development is due to a lack of sperm. When my colleague Nicola Hemmings and I started to examine the unhatched eggs from different species we found exactly the opposite. Of 5,975 great tit eggs, 11 per cent failed to hatch and of those almost all (98 per cent) had been fertilised. Similarly, of 7,813 blue tit eggs, 3.6 per cent failed to hatch and 97 per cent of those had been fertilised. These are common wild birds, so early embryo mortality, for whatever reason, is a regular occurrence, even in species whose populations are apparently healthy. When we extended our study to look at the unhatched eggs of endangered birds, the story was the same in most cases. For example, in the hihi, or stitchbird, of

New Zealand, whose surviving populations are largely confined to a small number of offshore, predator-free islands, 35 per cent of eggs failed to hatch even though 91 per cent of them had been fertilised. A shortage of sperm in this species would have been surprising given its highly promiscuous mating system, huge testes and frequent copulations.[9]

Early embryo mortality seems to be particularly common in birds with very small populations and the most likely cause is breeding with relatives. In other animals, including humans, spontaneous abortion is much more frequent when close relatives mate with each other.[10] It is precisely for this reason that most human cultures and religions have strong taboos against marrying close relatives. Even if such inbreeding does not result in spontaneous abortion and embryo death, it sometimes results in poor quality offspring. In Charles Darwin's day the taboo against marrying kin went only as far as first order relatives like brothers and sisters; it was perfectly acceptable to marry a cousin, as Darwin did. However, during Darwin's lifetime it became increasingly apparent that reproducing with a relative as close as a cousin might not be a good idea, and Darwin wondered whether his brood of sickly offspring – several of whom died in infancy – might have been partly due to their being the product of marriage between cousins.

Taking the last few remnants of a critically endangered bird population into captivity and attempting to breed them is often the only thing that can be done to try to save them. In some cases it seems to work well, as it has with Californian condors; in other cases hatching failure becomes even more common and breeding success even lower. Perhaps the most extreme example of hatching failure occurs in Spix's macaw, extinct in the wild and with a world population currently consisting of about seventy captive individuals.[11] Very few Spix's eggs hatch and when we looked at them we found that a substantial proportion – around half – had failed to be fertilised.

The reason seems obvious, for inbreeding in this captive population is extreme – virtually all of the birds alive today are derived from the same captive pair. Defective testes seems to be one of the costs of being inbred. Most of the remaining Spix's macaws are essentially clones of each other, with precious little genetic variation. Nonetheless, by examining their unhatched eggs and identifying which males did manage to get sufficient sperm to the female's ova, we have at least helped the programme to maximise its chances of saving this magnificent bird.[12]

As a biologist, I think of birds' eggs as examples of perfection – or at least the perfect compromise between the different selection pressures they experience. I happen to think of them as perfect, too, from an aesthetic point of view – in terms of colour, shape and size. These two perspectives aren't independent, of course: part of my biological fascination for eggs is driven by my admiration for their beauty.

How did such perfection evolve? There's insufficient space to recount the story of the evolution of eggs from the simplest forms of life, up to birds, but it is worth thinking about what happened during the most recent phase in the evolution of birds' eggs. After years of speculation, the evidence is now overwhelming that birds are dinosaurs. Like other reptiles, dinosaurs probably laid unmarked, symmetrical eggs (that is with no obvious blunt end), that were warmed in a nest either by heat from the sun or by heat from decomposing vegetation. It also seems that, like today's crocodilians, some dinosaurs provided parental care by guarding the nests in which their eggs were buried. What we do not know is when proper incubation – contact incubation – began. By this, I mean when dinosaurs started to use their own body heat to warm their eggs. Some researchers have maintained that dinosaurs were 'warm-blooded', but this isn't necessarily the same as being able generate heat that can be transferred to eggs. There's a great deal of heated discussion – if you'll forgive the pun – about this. Contact incubation may have evolved when certain lines of dinosaurs, protected by an insulating layer of feathers, were able to generate heat and maintain their own body temperature. It isn't a massive step from sitting on top of a nest to protect your eggs, to sitting on top of your eggs to give them some body heat. But as we saw in Chapter 6, contact incubation also necessitated a simultaneous change in egg design and composition. Unable to draw water into the egg from the surrounding vegetation or soil, as many reptile eggs do, bird embryos needed their own water supply and their eggs therefore contain a greater proportion of albumen than reptile eggs.[13]

Contact incubation has several advantages over reptilian incubation. First, it speeded things up – eggs heated by contact incubation hatch sooner than those warmed by the environment. Second, it freed birds from the unpredictability of the environment, allowing them to invade and utilise areas that were unsuitable for reptiles, and also to exploit areas already used by reptiles, but more efficiently. The ability to provide warmth for their eggs, together with other forms of paternal care, made bird reproduction more efficient and more successful than reptile reproduction. Success meant expansion – expansion geographically and ecologically. Just think about the range of situations that birds have evolved to cope with, from goldcrests incubating in cosy feather-lined nests in northern Europe, to emperor penguins breeding at temperatures of −50°C in the Antarctic winter, to grey gulls incubating in Chile's Atacama Desert where daytime temperatures exceed +50°C, grebes and divers incubating on waterlogged nests, megapodes leaving their eggs in piles of rotting vegetation, and guillemots incubating on bare but mucky cliff ledges with no nest.[14]

It is this incredible diversity of breeding environments that has created such a diversity of selection pressures on egg size, shape and colour, and it is these pressures that birds and their eggs have evolved in response to. As with all forms of life, what we see are the success stories: the adaptations that work. It is little wonder we think of eggs as perfect. What I find remarkable is that there has been enough genetic variation for natural selection to create so many ingenious compromises between the multitude of different selection pressures that different birds are subject to.

Perfection is relative. Birds' eggs are perfect only in the sense that they are the optimal compromise between different selection pressures. If those selection pressures change, what is perfect now may not be perfect in the future.

This is perhaps best demonstrated by the vast experiment unwittingly conducted by the poultry industry when they decided to incubate eggs artificially. Taking eggs away from their incubating parent was about as big a change as it was possible to make, as is apparent from how much research it took to fully understand what incubation comprises and to be able to recreate it artificially so as to produce perfect chicks.

Something similar may be happening with climate change. If the changes in environmental conditions are not too rapid, birds and their eggs may evolve to cope with the change. Birds have already evolved remarkable behavioural and physiological flexibility to deal with different incubation situations. It is not just that different species have evolved differences in egg design to cope with different circumstances; they have also evolved extraordinary physiological flexibility that allows, as we saw in Chapter 2, an individual female to produce eggshells of different design depending on whether she is breeding at sea level or partway up a mountain. Whether birds have the physiological wherewithal to produce eggs of different designs in response to changes in temperature, carbon dioxide levels or humidity as a result of climate change remains to be seen. But just as botanists have used plant specimens in museum collections (herbaria) collected in the last two centuries to track climate change through differences in the density of stomata in leaves, eggs in museum collections may also prove to be invaluable repositories of data for measuring climatic and other changes.[15]

And finally . . .

There are around five billion egg-laying hens in the world. China alone produces 490 billion eggs a year; the US 90 billion. This production has been achieved through selective breeding and careful environmental control so that, instead of laying a clutch of a dozen eggs like its wild ancestor, the red jungle fowl, commercial laying hens now produce around 300 eggs a year. Eggs are an important part of both our diet and our culture.

We eat a lot of hens' eggs: each of us in the UK eats around 200 a year – that is 11.5 billion altogether. They are cheap and nutritious and consumption has been encouraged by advertising, epitomised in the UK by the Egg Marketing Board's 1950s enduring catch-phrase 'go to work on an egg', suggesting that an egg for breakfast provides a good start to the day.[16]

Eggs are also symbols of fertility, more practical and whole-some than the equally important spermatozoa; somehow an 'Easter sperm' doesn't have quite the same appeal. In temperate regions of the world, birds – including farmyard fowl in the past – start to breed around Easter in response to increasing day length. As well as representing new life, eggs also symbolise rebirth, and for Christians the Easter egg also represents the resurrection of Christ. Catholics dyed hens' eggs red to signify the blood of Christ, the shell representing the tomb and the cracking of the shell the opening of that tomb. As with so many rituals the religious origins of Easter eggs have been obscured, partly by the widespread habit of decorating of hens' eggs, often in the most elaborate and beautiful ways, by the hunting for hidden eggs by children, and most of all by the mass production and consumption of chocolate eggs.

The other familiar egg image is Humpty Dumpty; the egg man, or the anthropomorphic egg. Emerging first in a riddle or nursery rhyme in the late 1700s, Humpty Dumpty falls from a wall, is smashed and cannot be put back together again. The rhyme's original meaning is lost but Humpty Dumpty may symbolise the fragility or vulnerability of man and the difficulty of reinventing oneself after being broken by a fall. Humpty Dumpty was resurrected in the 1870s by Lewis Carroll in *Through the Looking-Glass*, where, perched precariously on his narrow wall, he represents the worst kind of literary critic who through the use of jargon projects an impression of great profundity to intimidate the uninitiated normal person – represented in the book by Alice.[17]

I, too, have gone to work on an egg – a guillemot's egg – the extreme manifestation of extraordinary selection pressures. I have studied guillemots for forty years, motivated mainly by their vulnerability to marine pollution and trying to develop a robust understanding of their biology, including their eggs, so that they might be conserved. Less obviously conservable than puffins, which look so cute, guillemots are the avian mainstay of the marine ecosystem in the northern hemisphere. Their place in the marine food chain is pivotal. But tens of thousands die the most lingering and appalling death each year from oil pollution, and many more die as a result of overfishing and climate change. Guillemots are our barometers of marine wellbeing, and by failing to protect them – by continuing to overfish, by failing to prevent oil pollution, and by not doing enough about climate change – we are almost literally killing the goose that lays the golden egg.

After twenty-five years of continuous funding from the Welsh government the support for my long-term guillemot study was terminated in late 2013 because of budget cuts. The Skomer guillemot project was dead – one of many environmental casualties.

No sooner had the funding cuts been made when some of the worst storms in living memory – part of our changing climate – resulted in the death of at least 50,000 seabirds along Europe's western seaboard in the spring of 2014. The stormy conditions prevented the birds from feeding and they simply starved to death en masse. In what has become known as 'the seabird wreck', around half of the corpses were guillemots, and many of them were birds bearing Skomer rings and what I think of as *my* birds, in the sense that their conservation is my responsibility.

Understanding the consequences of this wreck is important. So I went back to the Welsh government, explained the situation and asked them if they would reconsider and reinstate the funding; I wrote articles and spoke to press, but all to no avail.

Luckily, at very short notice my university came to the rescue and provided the necessary funds to enable me to complete the 2014 field season and get a sense of what the wreck had done to the population. Not surprisingly, the main effect was extremely poor overwinter survival, with many fewer birds alive than in previous years. But it also resulted in many birds losing their lifelong partners, forcing them to re-pair and start the lengthy process of establishing a working partnership. For reasons like this, events as severe as seabird wrecks take several years to play out and studying that requires further funding.

Someone suggested I try to fund the research through crowd sourcing, but I wasn't optimistic. How would I ever make sufficient people aware, or find enough people who cared enough to make this work?

But several things conspired to make it happen. First, during the 2014 field season I had collaborated with a young visual artist, Chris Wallbank, to create huge images of guillemot colonies encompassing both my research and the spectacle of large numbers of birds on the cliffs. Chris and I hit it off immediately and I loved his art, which he made on huge rolls of paper like Chinese scrolls – large enough to capture the full sense of a bustling colony. The end result was an exhibition, part of my university's Festival of the Mind, in Sheffield's cathedral in September 2104. It would have been hard to find a more apposite venue – the impact of Chris's 'loomery scrolls', as they were called, elevated by the cathedral's cliff-like architecture.[18]

Our combined effort impressed the media, and I was invited to describe the study – and the loss of funding – in an article for the science journal *Nature*. What I wrote was a plea for the value of long-term studies in general, not just my guillemot study. There is overwhelming evidence that long-term studies are disproportionately productive. This success comes from a combination of deep knowledge, of researchers really knowing their study organisms and of the study having a long view – seeing organisms over a range of environmental conditions: good years, bad years – and ongoing climate change. Perhaps the single most important aspect of long-term studies is that they will allow us to investigate environmental problems that we haven't yet even imagined. Just as museum eggs collected for one purpose in the 1800s and 1900s were later used to look at the effects of acid rain and to establish pesticides as the cause of hatching failure in raptors (Chapter 2), long-term ecological studies are an investment in our environmental future.

Nature was a coup. It is the premier science journal with a worldwide readership, and I knew that this was the publicity that could launch the crowd sourcing. The combination of that article and contacting everyone I knew created a massive response. It was exhilarating and for two weeks my computer buzzed continuously with 'you have new mail' – new donations. I was thrilled, partly because I could see that the campaign was likely to be successful, but much more because of the messages I received from well-wishers recognising the value of environmental monitoring, of long-term studies and of beating the system. Officialdom might not care about the environment, but huge numbers of people do.[19]

For the time being, my study of guillemots, the birds and their eggs looks safe and set to continue. This book is dedicated to all of those who contributed to the campaign and made this possible.

Notes

PREFACE

1. Drane, R. (1897; 1898–9). The eggs were photographed and coloured by a Mr Charles E. T. Terry, and executed (printed, presumably) by Messers Bemrose & Sons, Derby; J. J. Neale was also closely involved in the Cardiff Naturalists, and twice its president. He died in 1919.

2. The image of Vaughan Palmer Davies's girls and their friend Ann Lush blowing guillemot and razorbill eggs on Skomer sometime between the 1860s and 1892 is from Howells (1987). Davies lived on Skomer between those dates, so this precedes Drane's visit. I checked whether the eggs Drane had so beautifully reproduced were still in 'his' Cardiff museum – but (as is so often the case, it seems) they are not. The current curator, Julian Carter, asked ex-curator Peter Howlett, but there is no record of these eggs.

3. Higginson (1862). The Life of Birds, *Atlantic Monthly* 10: 368–76.

1. CLIMMERS AND COLLECTORS

1. Vaughan (1998).

2. Whittaker (1997): George Rickaby's Bempton climming diary, found in a second-hand bookshop, was purchased and published by Whittaker.

3. Fowling, that is, the taking of eggs and adult seabirds, has occurred at Bempton since the 1500s: there is a record that William Strickland held the fowling rights there in the sixteenth century (Exchequer, King's Rembrancer Miscellaneous Book Series 1, 164/38 f.237).

4. Birkhead (1993).

5. Vaughan (1998).

6. J. Whittaker (personal communication, 21/02/2014).

7. Kightly (1984).

8. The photograph of Patricia Lupton was kindly provided by the Lupton family.

9. Whittaker (1997).

10. Cocker (2006). In North America collecting became illegal much earlier, in 1918.

11. For birds changing sex and males laying eggs, see Birkhead (2008).

12. John Evelyn's diary: http://www.gutenberg.org/ebooks/41218

13. Wood (1958).

14. Salmon (2000).

15. Newton (1896: 182). See also Cole (2016).

16. Cocker (2006).

17. One of the first people to think that eggs might provide insight into the arrangement of birds was Thienemann (1825–38); one recent molecular arrangement is by Jetz et al. (2012).

18. Newton (1896: 182–4).

19. Prynne (1963) refers to a 1959 television programme in which Sir Charles Wheeler, sculptor and president of the Royal Academy, spoke about the 'perfection of form', apparently drawing parallels between the ovals of eggs and of the female form. I have not been able to trace the television programme.

20. Manson-Bahr (1959).

21. One of the few artists to include birds' eggs in his images was Viscount Leroy de Barde (1777–1828). Sculpture: just Google 'egg

and sculpture'. My colleague, the artist Carry Akroyd (personal communication, November 2014), suggests that birds' eggs were simply not worth including in still-life paintings because they had no role in advertising anyone's wealth or status.

22. The state of Lupton's guillemot eggs at the Natural History Museum at Tring is, rather sadly, a consequence of the unusual way eggs were obtained at Bempton: huge numbers and poorly documented as far as dates and precise locations are concerned. Quite rightly, the Natural History Museum prioritises the curation of data-rich, scientifically valuable collections.

23. I searched in vain for the Metland eggs: twenty-six consecutive years of eggs from the same female! They may exist somewhere, for these must have been the pride of Lupton's collection. It is just possible they disappeared during the time between Lupton's death and his collection eventually finding a home in the museum at Tring.

24. B. Stokke (personal communication, 23/01/2014) extracted DNA from egg membranes; see also Green (1998), Green and Scharlemann (2003) and Russell et al. (2010) on the value of museum egg collections.

2. MAKING SHELLS

1. Rahn et al. (1979).

2. Gebhardt (1964).

3. Tyler (1964); Kutter was president of the DO-G (Deutsche Ornithologen-Gesellschaft, the German ornithological society) in 1890–1 (see Stresemann, 1975); some of Nathusius's shell sections are still in the Natural History Museum, Vienna.

4. Johnson (in Sturkie, 2000); see also Romanoff and Romanoff (1949: 144) on the thickness of two layers of shell membrane – the outer is slightly thicker.

5. Burley and Vadhera (1989: 58–9): the sequence isn't clear; it appears that initial calcification – the production of the mammillae (in the red region of the uterus) – precedes plumping.

6. Aristotle (*Generation*, book 3); William Harvey in Whitteridge (1981: 63): it is Harvey not Aristotle who refers to pushing an egg through the neck of a bottle.

7. Barn swallow information from Angela Turner, cited in Reynolds and Perrins (2010).

8. Reynolds and Perrins (2010).

9. Johnson (2000: 590). Konrad Lorenz (1965: 14) suggested that birds seeking calcium will peck at any white, hard, crumbling substance, which he says – in the ethological language of the day – acts as a 'special releasing mechanism'.

10. E. Roura (personal communication); Tordoff (2001).

11. Hellwald (1931).

12. Crossbills: Payne (1972) eating mortar, cited in Tordoff (2001). It turns out to be true: pine nuts contain just 1 per cent of calcium, and we don't know how much of that the birds can utilise; Jonathan Silvertown (personal communication, 30/09/14).

13. MacLean (1974).

14. Mongin and Sauveur (1974).

15. Graveland (1998: 45).

16. Reynolds and Perrins (2010).

17. Drent and Woldendorp (1989).

18. Graveland and Baerends (1997).

19. Green (1998). Interestingly, there was no indication that pesticides – which were responsible for a major reduction in shell thickness of birds of prey and fish-eating birds from the mid-1940s – had any effect on the thrushes.

20. Reynolds and Perrins (2010).

21. DDT is dichlorodiphenyltrichloroethane; DDE (dichlorodiphenyldichloroethylene) is its metabolite. Inspired by finding a rook's

eggshell, Mark Cocker in his book *Claxton* (2014) discusses the scourge of DDT.

22. Birkhead et al. (2014: 415) and http://www.nytimes.com/2010/11/16/science/16condors.html?pagewanted=all&_r=0

23. Carson (1962); see also *Merchants of Doubt* by Oreskes and Conway (2010). Neonicotinoids, another group of more recently developed pesticides, seem to be having the same devastating effect on wildlife. See Goulsen (2013).

24. Prynne (1963).

25. Grieve (1885: 104); Thienemann (1843).

26. Various methods have been devised for estimating the number of pores in eggshells but none of these could be used on the eggs of great auk.

27. Fuller (1999).

28. The chorio-allantois in birds' eggs serves the same purpose as the placenta in eutherian mammals.

29. Siegfried (2008). Although Davy is credited with the discovery of eggshell pores, Fabricius came close in the late 1500s when he said: 'the eggshell is porous (as is clear from the sweat which exudes from a fresh egg baked upon ashes)', although his interpretation was that the porosity is to facilitate heat transfer from the brooding hen to the embryo (Fabricius 1621: see Adelmann 1942: 215).

30. Davy (1863).

31. Romanoff and Romanoff (1949: 166–7).

32. Rahn, Paganelli and Ar (1987). The pioneer of eggshell thickness measurements was the German ornithologist and oologist Max Schoenwetter (1864–1961). His goal was to obtain specimens of the eggshells of all species of birds and to be able to deduce a number of features – including eggshell thickness, surface area and volume of the egg – from their length, breadth and weight. He succeeded and the end result of all this oological (and mathematical) endeavour was the massive *Handbook of Oology*, published in parts starting in 1960,

the year before Schoenwetter died. The remaining forty-six parts were edited and published over the next thirty years by Wilhelm Meise. It has been referred to as the 'great unread masterpiece of ornithology' (Maurer et al. 2010).

33. Sossinka (1982).

34. Rahn and Paganelli (1979; 1991). Hermann Rahn (1912–1990) is widely regarded as the most extraordinary, most ingenious and successful of egg biologists (Pappenheimer 1996).

35. Rahn et al. (1977, 1982).

36. Bertin et al. (2012).

3. THE SHAPE OF EGGS

1. Birkhead (1993).

2. Pyri: my *Collins English Dictionary* (1991) says 'erroneously from Latin *pirum*, pear'.

3. Newton (1896); Pitman (1964); Hauber (2014).

4. Bradfield (1951): in fact Geirsberg (1922) had also suggested this.

5. Warham (1990: 289).

6. Prynne (1963); Rensch (1947); C. Deeming (personal communication).

7. Newton (1896).

8. Cherry-Garrard (1922).

9. Hewitson (1831, vol. 2: xii).

10. Andersson (1978); see also Norton (1972), who suggested that a pointed egg shape reduces the rate of heat loss when the eggs are not incubated.

11. Harvey in Whitteridge (1981).

12. Debes is reprinted in Ray (1678). Although Ray and Willughby include a summary of Debes's accurate account of guillemot and seabird biology, they themselves have little to add, even though they

made a point of visiting seabird colonies around the British coast. Interestingly, they don't mention Harvey's egg-glued-to-a-rock idea.

13. Martin (1698). A university educated man, Martin was persuaded by a colleague to undertake a survey of Scotland's Western Isles 'more exactly than any other'. On St Kilda he found a contented community who, he said, consumed an extraordinary quantity of eggs. His *daily* maintenance provided by the residents while he was there consisted of barley cake and eighteen guillemot eggs.

14. Pennant (1768).

15. Blackburn (1987); Waterton (1835; 1871).

16. Waterton (1835; 1871).

17. Hewitson (1831).

18. Morris (1856); Allen (2010).

19. MacGillivray (1852).

20. Dresser (1871–1881, vol. 8: 753).

21. Uspenski (1958: 126): use of eggs in soap – the yolks were a source of fat.

22. Belopol'skii (1961: 6).

23. Belopol'skii (1907–1990); the ship is sometimes called *Chelyuskin*.

24. Wikipedia.org/wiki/ChukchI_Sea

25. Kaftanovski (1941a: 60; 1951).

26. Belopol'skii (1961: 130).

27. Belopol'skii (1961: 132).

28. Ibid.

29. Much has been written about Trofim Lysenko (1898–1976) and his appalling effect on both Soviet science and society. Adaptations, then and now: the key difference is that prior to the 1960s, Lysenko apart, much evolutionary thinking was in terms of group selection (adaptations favouring the group of species) rather than individual selection (favouring 'selfish' individuals).

30. Nowak's (2005) account is based on an interview with Belopol'skii in 1959 at the 2nd Ornithological All-Unions-Conference in Moscow.

31. Beat Tschanz (1920–2013): much of the information here was kindly provided by his student Paul Ingold (personal communication, 24/09/13).

32. Cullen (1957).

33. Tschanz (1990); additional information from Paul Ingold (personal communication, 24/09/13).

34. Tschanz et al. (1969) and Drent (1975).

35. Drent (1975: 372).

36. Geist (1986) showed that Bergmann's rule isn't true, but egg size in guillemots *Uria* spp. does increase with body size (Harris and Birkhead 1985: 168).

37. See Ingold (1980). We found that egg volume (equivalent to weight) explained less than 3 per cent of the variation in pointedness. The extent to which one variable explains variation in the other is referred to by statisticians as the 'coefficient of variation', and usually expressed as a percentage. The easiest way to think about this is to imagine the relationship between the height and weight of people: if weight was determined solely by a person's height, then all (100 per cent) of the variation in weight would be accounted for by height. But it isn't: the amount of body fat also has a huge effect on weight, so the relationship between height and weight has a large amount of scatter, and height explains only around 50 per cent of the variation in weight. Fifty per cent is still biologically meaningful, but less than 3 per cent isn't, and strongly suggests that one or more other factors are important.

38. Romanoff and Romanoff (1949: 262).

39. Romanoff and Romanoff (1949: 286).

40. Whittaker (1997).

41. Romanoff and Romanoff (1949: 280).

42. Fabricius in Adelmann (1942: 212).

43. Fabricius in Adelmann (1942: 212); Harvey in Whitteridge (1981: 310).

44. Barta and Szekely (1997).

45. Bain (1991).

4. COLOURING EGGS – HOW?

1. My tinamou hypothesis is a straw man: it is a device to make the point about different ways of looking at a biological problem. I do not really think that their eggs are unpalatable (although they might be). The issue of tinamou eggshell design is discussed by Igic et al. (2015).

2. Notably John Ray in the seventeenth century (Birkhead 2008).

3. Davies et al. (2012).

4. Higham (1963).

5. Previous researchers are listed and cited in Giersberg (1922) and include Tiedemann (1814) and later Wicke (1858), Blasius (1867), Ludwig (1874) and Liebermann (1878); whereas Carus (1826), Coste (1847) and Leuckart (1854) already assumed that the colouring took place in the uterus.

6. Opel (1858); Dresser (1871–881, vol. 8: 753).

7. Sorby (1875).

8. Ibid.

9. Battersby (1985).

10. In fact the blue-green ground pigment in the eggs of crows and thrushes had been discovered several years prior to Sorby's (1875) work by a German researcher Wicke (1858) and referred to as biliverdin.

11. Thomson (1923: 278) https://archive.org/details/biologyofbirdsoothom

12. Ibid.

13. Ibid.

14. Ibid. and note that Thomson (1923: 278) uses a plate of guillemot eggs as the frontispiece of his book. See also Hamilton and Brown (2001) for a discussion of the adaptive significance of autumn leaf colour.
15. Kennedy and Vevers (1976); Schmidt (1956).
16. Tyler (1969: 102).
17. I think he was wrong: my colleague Mike Harris examined a dead female guillemot with a fully formed, perfectly coloured egg in its uterus and told me the colour was fixed.
18. Kutter (1877–8); Taschenberg (1885); Romanoff and Romanoff (1949); Gilbert (1979).
19. Tamura and Fuji (1966); Lang and Wells (1987). Pike (2016).
20. Romanoff and Romanoff (1949: 100).
21. Ian Newton (personal communication).
22. These ideas are reviewed in Kilner (2006). The original running out of pigment idea was proposed by Nice (1937) and is supported to some extent by Lowther's (1988) observation that in the house sparrow unusually coloured last eggs are more likely in larger clutches.
23. Maurer et al. (2011). In a comparative study Maurer et al. (2015) found that the eggshells of cavity nesting birds allowed more light to penetrate the shell, even after taking eggshell thickness into account. They also suggest that eggshell colour may be a trade-off between allowing sufficient light in and keeping potentially harmful ultraviolet light out, away from the embryo. Comparative studies like these provide potential explanations for egg colouration, but what's needed now are experimental tests of these exciting ideas.
24. Cassey et al. (2011); Maurer et al. (2015).
25. Gaston and Nettleship (1981: 170).
26. Whittaker (1997).
27. Nathusius, cited in Tyler (1964: 590).
28. John Clare's 'The Yellowhammer's Nest': see: http://www.poetry-foundation.org/poem/179904
29. Whittaker (1997).

5. COLOURING EGGS – WHY?

1. http://darwin-online.org.uk/content/frameset?itemID=F350&view type=text&pageseq=1

2. Most often males compete for females and females choose between males, but occasionally it works the other way round. Among birds, phalaropes are an example: females are larger, brighter and more competitive than males.

3. Wallace A. R. (1895). *Natural selection and tropical nature*, 2nd edn, New York, NY: Macmillan and Co., pp. 378–9; see also: Prum: http://rstb.royalsocietypublishing.org/content/367/1600/2253. full#sec-1

4. The Darwin–Wallace debate is discussed by Cronin (1991).

5. Discussed by Birkhead (2012).

6. Erasmus Darwin (1794): http://quod.lib.umich.edu/e/ecco/0048 74881.0001.001/1:7.39.5?rgn=div3;view=fulltext p. 510; Charles Darwin (1875) does discuss eggs briefly in *Domestication*, repeating information from Dixon (1848).

7. See Kilner (2006); but also see Wiemann et al. (2015) who – remarkably – discovered blue-green pigmentation in the eggs of the sixty-six-million-year-old dinosaur *Heyuannia huangi* (an oviraptrid). The pigments in these fossil eggshells are the same two that occur in the eggshells of modern birds. The fact that *Heyuannia huangi* probably also incubated in a partially open nest suggests that open nesting and egg colouration evolved together.

8. There is no mention of Hewitson (1831) in Wallace's *Darwinism* (1889).

9. Wallace (1889: 214).

10. Thomson (1923).

11. McAldowie (1886): another explanation for white eggs among cavity nesting birds is that white eggs are more visible than pigmented eggs and therefore less likely to be trodden on and damaged by the parent birds – an idea first suggested by Henry Seebohm (1883).

12. Cassey et al. (2012) found several associations between egg colour and life history and breeding biology.

13. Cook et al. (2012); Webster et al. (2009).

14. Lovell et al. (2013).

15. Gosler et al. (2005); Magi et al. (2012).

16. Swynnerton (1916).

17. Cott (1951, 1952); see also Birkhead (2012).

18. Butcher and Miles (1995).

19. Moreno and Osorno (2003); Moreno et al. (2004, 2005).

20. Reynolds et al. (2009).

21. English and Montgomerie (2011).

22. Montevecchi (1976); Bertram and Burger (1981).

23. Gottlob Heinrich Kunz (1821–1911), Baldamus's colleague did, see Schulze-Hagen et al. (2009).

24. Cole and Trobe (2011: 22).

25. This is from John Robjent's unpublished diary.

26. In fact the BMNH at Tring subsequently obtained Robjent's diaries.

27. The cuckoo finch is a type of weaverbird and most closely related to another group of brood parasitic finches, the indigo birds and whydahs (Sorensen and Payne 2001).

28. Davies (2015).

29. These 'types', first noted in the common cuckoo in the 1800s, are referred to as gens (plural: gentes) and are now known to occur in several brood parasites.

30. See: http://www2.zoo.cam.ac.uk/africancuckoos/systems/cuckoo finch.html

31. Pennant (1768).

32. Gurney (1878) refers to a Sam Leng, who was a climmer (died in 1935 aged fifty-six – possibly his father or another relative).

33. Dresser (1871–81): I have been unable to identify any of Gray's publications that match what Dresser (vol. 8: 753) says, although we know that Gray visited and wrote about Ailsa Craig and its birds.

Dresser (vol. 8: 572) also comments on the widespread belief in Shetland that guillemots *deliberately* knock their eggs off the breeding edges when approached, rather than allow them to be taken by people collecting eggs.

34. Tuck (1961: 127).

35. Tschanz et al. (1969).

36. Gaston et al (1993).

37. Davies (2015).

38. Lyon (2003).

39. Lyon (2007).

40. Bertram (1992). Another group-nesting bird where recognition is important is the greater ani, a tropical cuckoo. Its unusual eggs – blue, with a white chalky bloom of Wedgwood-like vaterite – have long fascinated ornithologists and oologists. The lattice is designed to wear off as a result of abrasion in the nest and the nest-owning female uses the degree of wear to identify and eject eggs laid parasitically by other anis (Riehl 2010).

41. Cassey et al. (2011).

6. MUCH ADO ABOUT ALBUMEN: THE MICROBE WAR

1. Aristotle in Harvey in Whitteridge (1981: 173).

2. Harvey in Whitteridge (1981: 173), see also Adelmann (1942: 156).

3. Harvey in Whitteridge (1981: 174).

4. Harvey in Whitteridge (1981: 319).

5. Harvey in Whitteridge (1981: 475).

6. Giersberg (1922) says it was Giacomini in 1893. It was the poultry biologists Raymond Pearl and Maynie Curtis (1912) who recognised that the albumen secreted in the magnum region was much denser than that which eventually surrounds the yolk in the completed egg.

7. Haines and Moran (1940), cited in Romanoff and Romanoff (1949: 169); Gole et al. (2014).

8. There are over 100 species of salmonella, some of which are dangerous.

9. Board and Tranter (1995). Last sentence according to the *Daily Telegraph*, 26 December 2001.

10. http://www.forbes.com/sites/nadiaarumugam/2012/10/25/why-american-eggs-would-be-illegal-in-a-british-supermarket-and-vice-versa/

11. Van Wittich (cited in Romanoff and Romanoff 1949: 495).

12. Davy (1863).

13. There was another, rather different kind of academic described in Malcolm Bradbury's *The History Man* that one could also say epitomised that era.

14. Baudrimont and St Ange (1847).

15. This was first suggested by Nathusius (1884, 1887: see Tyler (1964)).

16. Board and Scott (1980) suggested the term SAM, which according to Sparks (1994) is made up of microscopic spheres (0.5–3μm in diameter).

17. Board (1981); Lack (1968).

18. Board (1981).

19. Board and Tranter (1991).

20. Wurtz (1890), cited in Romanoff and Romanoff (1949: 499).

21. Burley and Vadhera (1989: 295–6).

22. Board and Fuller (1974; 1994); see also Benrani et al. (2013). The yolk also contains antibodies, derived from the mother, that help to protect the developing embryo from infections.

23. Nys and Guyot (2011). Fabricius (1621) noted that albumen comprised different fractions (Adelmann 1942). The role of the different albumen fractions doesn't seem to be known (Y. Nys, personal communication, 10/12/14). One possibility is that four concentric

layers that clearly differ in their physical, and possibly chemical, properties make it harder for microbes to get to the embryo. The formation of the chalazae is discussed by Rahman et al. (2007).

24. Rahn (1991).

25. pH is measured on a log scale: acid = 1–2, 7 is neutral. Davy (1863). Board and Tratner (1995) suggest this but it is a guess.

26. Cook et al. (2005).

27. Very occasionally birds can lay two eggs in one day. Most tropical birds produce smaller clutches than temperate species.

28. Board and Fuller (1974).

29. Bessinger et al. (2005).

30. Ibid.

31. Ray (1678: 385).

32. It was also once believed that ostriches abandoned their eggs or buried them in the sand where they were incubated by the heat of the sun. There are numerous illustrations in medieval manuscripts – few of them resembling ostriches because few illustrators had ever seen one – of them doing so. See: http://bestiary.ca/beasts/beastgallery238.htm#

33. Board et al. (1982).

34. D'Alba et al. (2014).

35. Deeming (2004: 63, 262).

36. Board and Fuller (1974).

37. Romanoff and Romanoff (1949).

38. Orians and Janzen (1974).

39. Buffon (1770–83: vol. VII: 336–53).

40. Christian Ludwig Nitzsch (1782–1837), who was director of the Halle Museum in Germany, was later described by the great ornithologist Erwin Stresemann as 'one of the most accurate, cautious and imaginative morphologists who ever concerned themselves with avian anatomy' (Stresemann 1975: 308). Nitzsch's description of the hoopoe's preen gland appears in *System der Pterylographie*, published

with H. Burmeister in 1840, and translated into English in 1867 with the title *Nitzsch's Pterylography*.

41. *Bacillus licherniforis* – Soler et al. (2008).

42. Nathusius (1879), cited in Tyler (1964).

43. It is not known whether the hoopoe's albumen contains any special antimicrobial properties.

44. Vincze et al. (2013).

45. Walters (1994); Hauber (2014).

46. Hincke et al. (2010); Ishikawa et al. (2010) – cited in Cassey et al. (2011).

47. Cassey et al. (2012).

48. We have made a start on this . . .

49. Board and Fuller (1974); Kern et al. (1992).

50. Beetz (1916).

51. Hill (1993); Finkler et al. (1998). It is interesting that reptile eggs are much more amenable than birds' eggs to having some of their yolk removed – probably because their eggs contain much less albumen. In reptiles yolk removal results in smaller, less mobile hatchlings (Sinervo 1993). In birds the albumen is important and the ratio of albumen to yolk crucial. Experiments that remove only yolk upset the albumen–yolk balance, making it difficult to tell whether the effect on the resultant embryo is due to the reduction in yolk or change in the albumen–yolk balance (Finkler et al. 1998).

52. Carey (1996); Sotherland et al. (1990); Hill (1993); Romanoff and Romanoff (1949).

7. YOLK, OVARIES AND FERTILISATION

1. Kerridge (2014).

2. Ray (1678); the quote is from Pearl and Schoppe (1921). Birds are unusual among vertebrates in possessing only a single (left) ovary,

and the explanation is that this is a weight-saving device associated with flight. Kiwis are exceptional in having two functional ovaries.

3. Ray (1678: 10).

4. Abati (1589).

5. I asked Richard Serjeantson to look at Abati's original Latin text, and in his opinion it is not absolutely clear that Ray and Willughby's statement 'It is most probable, that hen birds have within them from their first formation all the eggs, they shall afterward lay throughout their entire lives', which they attribute to Abati, is what Abati says (R. Serjeantson, personal communication, 13/08/2013).

6. Karl Ernst von Baeyer in the 1820s was the first to see a human ovum.

7. Tilly et al. (2009).

8. Waldeyer (1870).

9. The quote is from Zuckerman (1965: 136).

10. Zuckerman (1965).

11. Ibid.

12. Wallace and Kelsey (2010).

13. Why so many sperm? The most compelling argument is 'sperm competition'. The females of most species are promiscuous and males compete for fertilisations; producing more sperm is the male's single best strategy for outcompeting the competition (Parker 1998).

14. Wallace and Kelsey (2010).

15. Tilly et al. (2011).

16. Willughby and Ray collected material for books on fishes and insects; their plans were disrupted when Willughby died in 1672, but Ray continued alone, seeing the *Ornithology* through the press, and writing and publishing *Historia Piscium* (fishes) in 1686 and *Historia Insectorum* (insects) in 1710.

17. Harvey in Whitteridge (1981: 154).

18. Ray (1678: 10).

19. Pearl and Schoppe (1921). It is ironic: but perhaps in birds it is a failsafe because birds have only a single functioning ovary, whereas women have two ovaries.

20. Harvey in Whitteridge (1981: 173–4).

21. Ibid.

22. Daddi (1896); Rogers (1908).

23. Grau (1976; 1982).

24. It is worth noting that potassium dichromate is both corrosive and carcinogenic. I say this because as a boy I was able to buy this and other chemicals from Reynolds and Branson's shop, not far from my school in Leeds, with no health warnings or restrictions. Grau (1976).

25. Roudybush et al. (1979); Astheimer and Grau (1990).

26. Birkhead and del Nevo (1987); Hatchwell and Pellatt (1990).

27. Burley (1988).

28. Cunningham and Russell (2000); Horvathova et al. (2011).

29. Aristotle: *History of Animals*, vol. vi: 2, 559a 20. In fact, the yolk is not perfectly spherical, as Bartelmez (1918) was at pains to point out. If you cut a window in an eggshell, with the pointed end of the eggshell lying towards the right, the yolk is longer from left to right than it is from top to bottom; it is also flatter on the top where the embryo lies, and the embryo is usually orientated such that it lies obliquely with its head at about two o'clock.

30. Tarschanoff (1884); Sotherland et al. (1990).

31. Schwabl (1993).

32. Aslam et al. (2013).

33. Gil et al. (1999).

34. Lipar et al. (1999).

35. Deeming and Pike (2013).

36. Short (2003); Harvey in Whitteridge (1981).

37. The frontispiece of Harvey's book shows Zeus with two halves of an egg releasing different sorts of animals. *De Generatione Animalium*

was first published in 1651 in Latin, with an English edition (*On the Generation of Animals*) in 1653.

38. Briggs and Wessel (2006); several people before Hertwig looked at the fertilisation of sea urchin eggs.

39. Alexander and Noonan (1979), Strassmann (2013).

40. Stepinska and Bakst (2007).

41. Astheimer et al. (1985).

42. The idea of a honeymoon period seems to have been used first with reference to the short-tailed shearwater by Marshall and Serventy (1956 – cited in Warham 1990) where, as in some other petrel species, both sexes do disappear together. In others, only the female departs, as summarised by Warham (1990: 258–60). Grey-faced petrel: Imber (1976 – cited in Warham 1990).

43. Snook et al. (2011).

44. Harper (1904).

45. Bennison et al. (2015).

46. But its success is almost certainly exaggerated. A fertile egg has sperm on the perivitelline layer (PVL) surrounding the yolk, holes in the PVL where sperm have penetrated, and embryonic cells in the germinal disc; an infertile egg has no (or few) sperm or holes and no embryonic cells in the germinal disc. Just to add a bit of confusion, zebra finch eggs sometimes start to develop without any male contribution, a phenomenon called 'parthenogenesis' – in these cases there are embryonic cells in the germinal disc, but there are no sperm or holes whatsoever (Birkhead et al. 2008).

47. Hemmings and Birkhead (2015).

8. STUPENDIOUS LOVE:
LAYING, INCUBATION AND HATCHING

1. Swift (1726). *Gulliver's Travels*.

2. Aristotle (*Generation*); Wirsling and Günther (1772); Meckel von Hemsbach (1851); Thompson (1908; 1917); Thomson (1923).

3. Purkinje (1830); von Baer (1828–88); Wickmann (1896).

4. Wickmann (1896).

5. Bradfield (1951).

6. Wickmann (1896); Sykes (1953).

7. Personal observation: incubation is shared between the partners in eight- to seventeen-hour stints for thirty-two days. I thank my field assistants Jodie Crane and Julie Riordan for video-recording egg laying so I could watch in detail.

8. Michael Harris (personal communication, 03/10/2014). It does not actually clinch it – what we really need is a bird with an incompletely formed egg.

9. Geese and ducks: Salamon and Kent (2014); Warham (1990: 273–4) says that petrel eggs are laid pointed end first and he cannot imagine the relatively large egg in small petrels turning before laying.

10. Emperor and king penguins – André Ancel (personal communication, 09/10/2014). Emperor and king penguins: this might be because their eggs are among the smallest, relative to body size, of any birds' eggs: for the emperor penguin just 2.3 per cent of female body weight (Williams 1995: 23). This is a far cry from the guillemot's 12 per cent. It may therefore be easier for an emperor penguin to lay an egg. Interestingly, emperor and king penguin eggs are laid on to the female's feet and not on the ice or the ground. This may explain why guillemot eggs are laid pointed end first – this is the thickest, strongest part of the egg: it allows the guillemot to have a thick-shelled egg to protect it from the substrate, but weak enough at the blunt end to allow the chick to get out. Waders: I was sent videos of a lapwing and an avocet laying, by Ingvar Byrkjedal and Astrid Kant, respectively, and in both cases the pointed end emerged first.

11. Hervieux de Chanteloup (1713).

12. According to Ord (1836), Alexander Wilson knew about early morning laying; yellow warbler: McMaster et al. (1999).

13. Davies (2015).

14. Ibid.: cuckoo eggs are laid at forty-eight-hour intervals and are fully formed after twenty-four hours.

15. Ray (1691) used the expression 'impugnable appetite' referring to the male's enthusiasm for copulation. In the case of the guillemot, as with many other species, it is to maximise their reproductive success.

16. Birkhead et al. (1985).

17. Livia was first married in 43 BC to Tiberius Claudius Nero (not the notorious Nero who later became emperor) and it was that marriage that this anecdote relates to. She had two sons by him. She was later married to the Emperor Augustus with whom she stayed married for fifty-one years, but this was a childless union apart from a miscarriage (J. Mynott, personal communication, 17/10/2014).

18. Réaumur (1750 – English edition: first published 1722).

19. Gaddis (1955).

20. Cantelo: Vanessa Toulmin (personal communication, 23/10/2014), 'Cantelo's Patent Hydro-Incubating Chicken Machine'. Overview of incubators: http://www.theodora.com/encyclopedia/i/incubation.html

21. Drent (1975); it is worth noting, too, that after a few days of development embryos generate some of their own heat as well.

22. Thaler (1990).

23. Ray (1678).

24. Boersma and Wheelwright (1979).

25. Birkhead (2008: 69).

26. Double-banded courser (Drent 1975); Egyptian plover (Howell 1979): interestingly, Howell found no evidence for any special adaptations in the plover's eggshell.

27. Ar and Sidis (2002); Deeming (2002: Chapter 10).

28. Réaumur (1750). The idea that unturned eggs fail because the embryo adheres to the shell membrane was started by Dareste (1891). For subsequent research see: Deeming (2002) and Baggot et al. (2002). C. Deeming (personal communication, 12/12/14).

29. Drent (1975).

30. Deeming (2002); relative amount of albumen in birds' eggs: Sotherland and Rahn (1987); in reptile eggs: Deeming and Unwin (2004).

31. Heinroth – see Schulze-Hagen and Birkhead (2009).

32. Lack (1968).

33. Burton and Tullett (1985).

34. Yarrell (1826).

35. Thomson (1964); Garcia (2007).

36. Bond et al. (1988); Adelmann (1942: 224).

37. Garcia (2007).

38. Bond et al. (1988).

39. Ibid.; Tschanz (1968).

40. Tschanz (1968).

41. Heinroth (1922, cited in Drent 1975); Schulze-Hagen and Birkhead (2015: 40); Faust (1960).

42. Vince (1969).

43. Tinbergen et al. (1962).

9. EPILOGUE: LUPTON'S LEGACY

1. Louis Blériot had flown twenty-three miles across the English Channel in 1909. In fact Hewitt was *second* to fly across the Irish Sea (Dennis Wilson flew from Goodwick in Pembrokeshire to Enniscorthy in Ireland four days earlier), but Hewitt's flight was considered the more dangerous of the two. He didn't get quite the recognition some thought he deserved because news of his success was eclipsed by the *Titanic* disaster just two weeks earlier.

2. Howells (1987); Hewitt information from David Wilson and Jeremy Greenwood (personal communication, 17/03/2014). After Hewitt's death 'There were trunks full of guillemot eggs, trunks full of razor-bill eggs' (David Wilson, personal communication, 17/03/2014).

3. Jim Flegg and Jeremy Greenwood (personal communication, 01/02/2014).

4. Douglas Russell (personal communication, 12/02/2014).

5. Jean E. Woods (personal communication, 20/02/2014): 'Unfortunately this part of our egg collection is poorly curated and quite extensive. The biggest challenge is that the data slips were shipped to us separately from the eggs themselves and have mostly not been re-matched. I've had a look through the eggs and while I find various pieces of paper with Lupton's name (e.g. ex Lupton Coll.) on them I can't really match them to any particular eggs. Nor was I lucky enough to find the eggs in the subset where the data slips are with the eggs.' Note: to add a gloss of complexity to the whole business, du Pont was convicted of murdering a friend in 1997.

6. This information from the Lupton family.

7. Tim Birkhead, Jamie Thompson and John Biggins, unpublished – so far.

8. Karl Schulze-Hagen (personal communication).

9. Hemmings et al. (2012).

10. http://www.nlm.nih.gov/medlineplus/ency/article/001488.htm

11. Juniper (2002).

12. Hammer and Watson (2012); Hemmings et al. (2012).

13. Deeming 2002 (in Deeming 2002); Deeming and Unwin (2004); Deeming and Ruta (2014).

14. Carey (2002) in Deeming 2002: 238–53.

15. McElwain and Chaloner (1996).

16. http://www.egginfo.co.uk/industry-data

17. Lewis Carroll, *Alice Through the Looking-Glass*; Priestley (1921).

18. A 'loomery' is an old term for a guillemot colony. Loomery Scrolls video – on YouTube – https://itunes.apple.com/gb/itunes-u/loomery-scrolls/id953108274?mt=10

19. *Nature*: http://www.nature.com/news/stormy-outlook-for-long-term-ecology-studies-1.16185; *Guardian*: http://www.theguardian.com/environment/2014/oct/26/guillemots-study-skomer-wales-budget-cut-tim-birkhead http://www.theguardian.com/science/animal-magic/2014/nov/10/crowdsourcing-funding-seabird-guillemot-skomer

Bibliography

Abati, B. A. (1589). *De admirabili viperae natura, et de mirificis eiusdem facultatibus liber.* Urbino.

Adelmann, H. B. (1942). *The Embryological Treatises of Hieronymus Fabricius of Aquapendente: The Formation of the Egg and Chick and the Formed Fetus.* Ithaca: Cornell University Press.

Alexander, R. D., and Noonan, K. M. (1979). Concealment of ovulation, paternal care, and human social evolution, pp. 436–53, in: *Evolutionary Biology and Human Social Behavior* (eds N. A. Chagnon and W. Irons). Belmont, Calif.: Duxbury Press.

Allen, D. E. (2010). *Books and Naturalists.* London: Collins.

Andersson, M. (1978). Optimal egg shape in waders. *Ornis Fennica* 55: 105–9.

Ar, A., and Sidis, Y. (2002). Nest microclimate during incubation, pp. 143–60, in: *Avian Incubation: Behaviour, Environment and Evolution* (ed. D. C. Deeming). Oxford: Oxford University Press.

Aristotle. *History of Animals.*

Aristotle. *On the Generation of Animals.*

Aslam, M. A., Hulst, M., Hoving-Bolink, R. A., Smits, M. A., de Vries, B., Weites, I., Groothuis, T. G., and Woelders, H. (2013). Yolk concentrations of hormones and glucose and egg weight and egg dimensions in unincubated chicken eggs, in relation to egg sex and hen body weight. *General and Comparative Endocrinology* 187: 15–22.

Astheimer, L. B., and Grau, C. R. (1990). A comparison of yolk growth rates in seabird eggs. *Ibis* 132: 380–94.

Astheimer, L. B., Prince, P. A., and Grau, C. R. (1985). Egg formation and the pre-laying period of black-browed and grey-headed albatrosses *Diomedea melanophris* and *D. chrysostoma* at Bird Island, South Georgia. *Ibis* 127: 523–9.

Baer, v. C. E. (1828–1888). *Ueber die Entwickelungsgeschichte der Thiere – Beobachtung und Reflexion*. I. Theil; 1828; II. Theil, 1837; II. Theil, 1888 (ed. L. Stieda). Leipzig.

Baggot, G. K., Deeming, D. C., and Latter, G. V. (2002). Electrolyte and water balance of the early avian embryo: effects of egg turning. *Avian and Poultry Biology Reviews* 13: 105–19.

Bain, M. M. (1991). A reinterpretation of eggshell strength, pp. 131–45, in: *Egg and Eggshell Quality* (ed. S. E. Solomon). Aylesbury: Wolfe Publishing.

Barta, Z., and Szekely, T. (1997). The optimal shape of avian eggs. *Functional Ecology* 11: 656–62.

Bartelmez, G. W. (1918). The relation of the embryo to the principal axis of symmetry in the bird's egg. *Biological Bulletin* 35: 319–61.

Battersby, A. R. (1985). Biosynthesis of the pigments of life. *Proceedings of the Royal Society of London B* 225: 1–26.

Baudrimont, A., and St Ange, M. (1847). Recherches sur les phenomenes chimiques de revolution embryonnaire des oiseaux et des bactraciens. *Annales de chimie et de physique* 21: 195–295.

Beetz, J. (1916). Notes on the eider. *Auk* 55: 387–400.

Beissinger, S. R., Cook, M. I., and Arendt, W. J. (2005). The shelf life of bird eggs: testing egg viability using a tropical climate gradient. *Ecology* 86: 2164–75.

Belopol'skii, L. O. (1961). *Ecology of Sea Colony Birds of the Barents Sea*. Jerusalem: Israel Program for Scientific Translations.

Bennison, C., Hemmings, N., Slate, J., and Birkhead, T. R. (2015). Long sperm fertilise more eggs in a bird. *Proceedings of the Royal Society of London B* 20141897. http://dx.doi.org/10.1098/rspb.2014.1897

Benrani, L., Helloin, E., Guyot, N., Rehault-Godbert, S., and Nys, Y. 2013. Passive maternal exposure to environmental microbes selectively modulates the innate defences of chicken egg white by increasing some of its antibacterial activities. *BMC Microbiology* 13: 128. doi:10.1186/1471-2180-13-128

Bertin, A., Calandreau, L., Arnould, C., and Levy, F. (2012). The developmental stage of chicken embryos modulates the impact of *in ovo* olfactory stimulation on food preferences. *Chemical Senses* 37: 253–61.

Bertram, B. C. R. (1992). *The Ostrich Communal Nesting System*. Princeton: Princeton University Press.

Bertram, B. C. R., and Burger, A. E. (1981). Are ostrich eggs the wrong colour? *Ibis* 123: 207–10.

Birkhead, T. R. (1993). *Great Auk Islands: A Field Biologist in the Arctic*. London: Poyser.

Birkhead, T. R. (2008). *The Wisdom of Birds*. London: Bloomsbury.

Birkhead, T. R. (2012). *Bird Sense: What It's Like to Be a Bird*. London: Bloomsbury.

Birkhead, T.R., Hall, J., Schutt, E., and Hemmings, N. (2008). Unhatched eggs; methods for discriminating between infertility and early embryo mortality. *Ibis* 150: 508–17.

Birkhead, T. R., Johnson, S. D., and Nettleship, D. N. (1985). Extra-pair matings and mate guarding in the common murre *Uria aalge*. *Animal Behaviour* 33: 608–19.

Birkhead, T. R., and del Nevo, A. (1987). Egg formation and the prelaying period of the common guillemot *Uria aalge*. *Journal of Zoology* 211: 83–8.

Birkhead, T. R., Wimpenny, J., and Montgomerie, R. (2014). *Ten Thousand Birds: Ornithology Since Darwin*. Princeton: Princeton University Press.

Blackburn, J. (1989). *Charles Waterton, 1782–1865: Traveller and Conservationist*. London: Bodley Head.

Board, R. G. (1981). The microstructure of avian eggshells, adaptive significance and practical implications in aviculture. *Wildfowl* 32: 132–6.

Board, R. G., and Fuller, R. (1974). Non-specific antimicrobial defences of the avian egg, embryo and neonate. *Biological Reviews* 49: 15–49.

Board, R. G., Perrott, H. R., Love, G., and Seymour, R. S. (1982). A novel pore system in the eggshells of the malleefowl *Leipoa ocellata*. *Journal of Experimental Zoology* 220: 131–4.

Board, R.G., and Scott, V. D. (1980). Porosity of the avian eggshell. *American Zoologist* 20: 339–49.

Board, R. G., and Tranter, H. S. (1994). The microbiology of eggs. Chapter 5 in: *Egg Science and Technology* (eds W. J. Stadelman and O. J. Coterill). New York: Food Products Press.

Boersma, P. D., and Wheelwright, N. T. (1979). Egg neglect in the Procellariiformes: reproductive adaptations in the fork-tailed storm-petrel. *Condor* 81: 157–65.

Bond, G. M., Board, R. G., and Scott, V. D. (1988). An account of the hatching strategies of birds. *Biological Reviews* 63: 395–415.

Bradfield, J. R. G. (1951). Radiographic studies on the formation of the hen's eggshell. *Journal of Experimental Biology* 28: 125–40.

Briggs, E., and Wessel, G. M. (2006). In the beginning . . . animal fertilization and sea urchin development. *Developmental Biology* 300: 15–26.

Buffon, G. L. (1770–83). *Histoire Naturelle des Oiseaux*. Paris.

Burley, N. (1988). The differential allocation hypothesis: an experimental test. *American Naturalist* 132: 611–28.

Burley, R. W., and Vadhera, D. V. (1989). *The Avian Egg, Chemistry and Biology*. New York: John Wiley & Sons.

Burton, F. G., and Tullett, S. G. (1985). Respiration of avian embryos. *Comparative Biochemistry and Physiology* 82A: 735–44.

Butcher, G. D., and Miles, R. D. (1995). Factors causing poor pigment of brown-shelled eggs. Cooperative Extension Service Fact Sheet VM94. Institute for Food and Agricultural Sciences, University of Florida, Gainesville, Fla.

Carey, C. (1996). Female reproductive energetics. In: Carey, C. (ed.) *Avian Energetics and Nutritional Ecology*. New York: Chapman & Hall.

Carson, R. (1962). *Silent Spring*. Cambridge, Mass.: Houghton Mifflin.

Cassey, P., Golo, M., Lovell, P. G., and Hanley, D. (2011). Conspicuous eggs and colourful hypotheses: testing the role of multiple influences on avian eggshell appearance. *Avian Biology Research* 4: 185–95.

Cassey, P., Thomas, G. H., Portugal, S. J., Maurer, G., Hauber, M. E., Grim, T., Lovell, G., and Miksik, I. (2102). Why are birds' eggs colourful? Eggshell pigments co-vary with life-history and nesting ecology among British breeding non-passerine birds. *Biological Journal of the Linnean Society* 106: 657–72.

Cherry-Garrard, A. (1922). *The Worst Journey in the World*. London: Carroll & Graf.

Cocker, M. (2006). End of the naturalists. *Guardian* 8 November 2006.

Cocker, M. (2014). *Claxton: Field Notes from a Small Planet*. London: Jonathan Cape.

Cole, E. (2016). Blown out: the science and enthusiasm of egg collecting in the *Oologists' Record*, 1921–69. *Journal of Historical Geography* 51: 18–28.

Cole, A. C., and Trobe, W. M. (2000). *The Egg Collectors of Great Britain and Ireland*. Leeds: Peregrine Books.

Cole, A. C. and Trobe, W. M. (2011). *The Egg Collectors of Great Britain and Ireland: An Update*. Leeds: Peregrine Books.

Cook, L. M., Grant, B. S., Saccheri, I. J., and Mallet, J. (2012). Selective bird predation on the peppered moth: the last experiment of Michael Majerus. *Biology Letters* 8: 609–12.

Cook, M. I., Bessinger, S. R., Toranzos, G. A., Rodriguez, R., and Arendt, W. J. (2005). Microbial infection affects egg viability and incubation behavior in as tropical passerine. *Behavioral Ecology* 16: 30–6.

Cott, H. B. (1951). The Palatability of the Eggs of Birds: Illustrated by experiments on the food preferences of the hedgehog (*Erinaceus Europaeus*). *Proceedings of the Zoological Society of London* 121: 1–41.

Cott, H. B. (1952). The Palatability of the Eggs of Birds: Illustrated by experiments on the food preferences of the hedgehog (*Erinaceus Europaeus*). *Proceedings of the Zoological Society of London* 122: 1–54.

Cronin, H. (1991). *The Ant and the Peacock: Altruism and Sexual Selection from Darwin to Today*. Cambridge: Cambridge University Press.

Cullen, E. (1957). Adaptations in the kittiwake to cliff nesting. *Ibis* 99: 275–302.

Cunningham, E., and Russell, A. (2000). Egg investment is influenced by male attractiveness in the mallard. *Nature* 404: 74–7.

D'Alba, L., Jones, D. N., Badway, H. T., Eliason, C. M., and Shawkey, M. D. (2014). Antimicrobial properties of a nanostructured eggshell from a compost-nesting bird. *Journal of Experimental Biology* 217: 1116–21.

Daddi, L. (1896). Nouvelle méthode pour colorer la graisse dans les tissus. In: *Archives Italiennes de Biologie* Tome 26: 143–6.

Dareste, C. (1891). *Recherches Sur La Production Artificielle Des Monstruosités Ou Essais De Tératogénie Expérimentale*, 2nd edn. Paris.

Darwin, C. (1875). *The Variation of Animals and Plants under Domestication*. London: John Murray.

Darwin, E. (1794). *Zoonomia, or the Laws of Organic Life*. London: Johnson.

Davies, N. B. (2015). *Cuckoo: Cheating by Nature*. London: Bloomsbury.

Davies, N. B., Krebs, J. R., and West, S. A. (2012). *An Introduction to Behavioural Ecology*. Oxford: Wiley-Blackwell.

Davy, J. (1863). Some observations on the eggs of birds. *Edinburgh New Philosophical Journal*, Series 2 18: 249–58.

Deeming, D. C. (ed.) (2002). *Avian Incubation: Behaviour, Environment and Evolution*. Oxford: Oxford University Press.

Deeming, D. C. (2004). *Reptilian Incubation*. Nottingham: Nottingham University Press.

Deeming, D. C., and Pike, T. W. (2013). Embryonic growth and antioxidant provision in avian eggs. *Biology Letters* 9: 20130757. http://dx.doi.org/10.1098/rsbl.2013.07577.

Deeming, C. and Ruta, M. (2014). Egg shape changes at the theropod-bird transition, and a morphometric study of amniote eggs. *Royal Society Open Science*. DOI: 10.1098/rsos.140311

Deeming, D. C., and Unwin, D. M. (2004), pp. 1–14, in: Reptilian incubation: evolution and the fossil record. In: Deeming, D. C. (ed.), *Reptilian Incubation*. Nottingham: Nottingham University Press.

Dixon, E. S. (1848). *Ornamental and Domestic Poultry*. London: Gardeners' Chronicle.

Drane, R. (1897). A Pilgrimage to Golgotha, June 1897. *Cardiff Naturalists' Society, Report & Transactions* 31: 38–51.

Drane, R. (1898–9). Eggs of the common guillemot and razorbill. *Cardiff Naturalists' Society, Report & Transactions* 31: 52–3.

Drent, R. (1975). Incubation. In: *Avian Biology* vol. V: 333–420 (eds D. S. Farner., J. R. King and K. C. Parkes). New York: Academic Press.

Drent, R., and Woldendorp, J. W. (1989). Acid rain and eggshells. *Nature* 339: 431.

Dresser, H. E. (1871–82). *A History of the Birds of Europe*. London: The Author.

English, P. A., and Montgomerie, R. (2011). Robin's egg blue: does egg color influence male parental care? *Behavioral Ecology and Sociobiology* 65: 1029–36.

Faust, R. (1960). Brutbiologie des Nandus (*Rhea americana*) in Gefangenschaft. *Verhandlungen der Deutschen Zoologischen Gesellschaft* 42: 398–401.

Finkler, M. S., Orman, J. B. van, and Sotherland, P. R. (1998). Experimental manipulation of egg quality in chickens: influence of albumen and yolk on the size and body composition of near-term embryos in a precocial bird. *Journal of Comparative Physiology B* 168: 17–24.

Fuller, E. (1999). *The Great Auk*. Southborough: Errol Fuller.

Gaddis, T. E. (1955). *Birdman of Alcatraz*. New York: Aeonian Press.

Garcia, R. A. (2007). An 'egg-tooth'-like structure in titanosaurian sauropod embryos. *Journal of Vertebrate Paleontology* 27: 247–52.

Gaston, A. J., Forest, L. de, and Noble, D. G. (1993). Egg recognition and egg stealing in murres (*Uria* spp.). *Animal Behaviour* 45: 301–6.

Gaston, A. J., and Nettleship, N. N. (1981). *The Thick-billed Murres of Prince Leopold Island*. Ottawa: Canadian Wildlife Service.

Gebhardt, L. (1964). *Die Ornithologen Europas*. Gießen: Brühlscher.

Geist, V. (1986). Bergmann's rule is invalid. *Canadian Journal of Zoology* 65: 1035–8.

Giersberg, H. (1922). Untersuchungen über Physiologie und Histologie des Eileiters der Reptilien und Vogel; nebst einem Beitrag zur Fasergenese. *Zeitschrift für wissenschaftliche Zoologie* 120: 1–97.

Gil, D., Graves, J., Hazon, N., and Wells, A. (1999). Male attractiveness and differential testosterone investment in zebra finch eggs. *Science* 286: 126–8.

Gilbert, A. B. (1979). Female genital organs. In: *Form and Function in Birds*, vol. 1: 237–360 (eds A. S. King and J. McLelland). London: Academic Press.

Gole, V. C., Roberts, J. R., Sexton, M., May, D., Kiermeier, A., and Chousalkar, K. K. (2014). Effect of egg washing and correlation between cuticle and egg penetration by various *Salmonella* strains. *International Journal of Food Microbiology* 182–183: 18–25.

Gosler, A. G., Higham, J. P., and Reynolds, S. J. (2005). Why are birds' eggs speckled? *Ecology Letters* 8: 1105–13.

Goulsen, D. (2013). An overview of the environmental risks posed by neonicotinoid insecticide. *Journal of Applied Ecology* 50: 977–87.

Grau, C. R. (1976). Ring structure of avian egg yolk. *Poultry Science* 55: 1418–22.

Grau, C. R. (1982). Egg formation in Fiordland crested penguins (*Eudyptes pachyrhynchus*). *Condor* 84: 172–7.

Graveland, J. (1998). Effects of acid rain on bird populations. *Environmental Reviews* 6: 41–54.

Graveland, J., and Baerends, J. E. (1997). Timing of the calcium uptake and effect of calcium deficiency on behaviour and egg-laying in captive Great Tits, *Parus major*. *Physiological Zoology* 70: 74–84.

Green, R. E. (1998). Long-term decline in the thickness of eggshells of thrushes, *Turdus* spp. in Britain. *Proceedings of the Royal Society of London B* 265: 679–84.

Green, R. E., and Scharlemann, J. P. (2003). Egg and skin collections as a resource for long-term ecological studies. *Bulletin of the British Ornithologists' Club* 123A: 165–76.

Grieve, S. (1885). *The Great Auk or Garefowl* Alca impennnis, *its History, Archaeology and Remains*. London: Thomas C. Jack.

Gurney, J. H. (1878). On Flamborough Head. In: *Ornithological Miscellany*, vol. 3: 29–38 (ed. Rowley, G. D.) London: Trubner & Co.

Hamilton, W. D., and Brown, S. P. (2001). Autumn leaf colour and herbivore defence. *Proceedings of the Royal Society of London B* 268: 1489–93.

Hammer, S., and Watson, R. (2012). The challenge of managing Spix macaws (*Cyanopsitta spixii*) at Qatar – an eleven year retrospection. *Der Zoologische Garten (Neue Folge)* 81: 81–95.

Harper, E. H. (1904). The fertilization and early development of the pigeon's egg. *American Journal of Anatomy* 3: 349–386.

Harris, M. P., and Birkhead, T. R. (1985). Breeding ecology of the Atlantic Alcidae, pp. 155–204, in *The Atlantic Alcidae* (eds D. N. Nettleship and T. R. Birkhead). London: Academic Press.

Hatchwell, B. J., and Pellatt, E. J. (1990). Intraspecific variation in egg composition and yolk formation in the common guillemot (*Uria aalge*). *Journal of Zoology, London* 220: 279–86.

Hauber, M. (2014). *The Egg Book*. Chicago: University of Chicago Press.

Heinroth, O. (1938). *Aus dem Leben der Vögel*. Berlin: Springer.

Heinroth, O., and Heinroth, M. (1924–34). *Die Vögel Mitteleuropas*, 4 vols. Berlin: Bermühler.

Hellwald, H. (1931). Untersuchungen über triebstärken bei tieren. *Zeitschrift für Psychologie und Physiologie der Sinnesorgane* 123: 94–103.

Hemmings, N., West, M., and Birkhead, T. R. (2012). Causes of hatching failure in endangered birds. *Biology Letters* 8: 964–7.

Hemmings, N. and Birkhead, T. R. (2015). Polyspermy in birds: sperm numbers and embryo survival. *Proceedings of the Royal Society of London B* 282: 000–000. http://dx.doi.org/10.1098/rspb.2015.1682

Hervieux de Chanteloup, J.-C. (1713). *Nouveau Traité des Serins de Canarie*. Paris: Claude Prudhomme.

Hewitson, W. C. (1831). *British oology: being illustrations of the eggs of British birds, with figures of each species, as far as practicable, drawn and coloured from nature: accompanied by descriptions of the materials and situation of their nests, number of eggs*. Newcastle upon Tyne: Empson.

Higginson, T. W. (1862). The life of birds. *Atlantic Monthly* 10: 36876.

Higham, N. (1963). *A Very Scientific Gentleman: The Major Achievements of Henry Clifton Sorby*. London: Pergamon Press.

Hill, W. L. (1993). Importance of prenatal nutrition to the development of a precocial chick. *Developmental Psychobiology* 26: 237–49.

Hincke, M. T., Nys, Y., and Gautron, J. (2010). The role of matrix proteins in eggshell formation. *Japan Poultry Science* 47: 208–19.

Horvathova, T., Nakagawa, S., and Uller, T. (2011). Strategic female reproductive investment in response to male attractiveness in birds. *Proceedings of the Royal Society of London B* 279: 163–70.

Howell, T. R. (1979). Breeding biology of the Egyptian plover, *Pluvianus Aegyptius*. University of California, Publications in Zoology.

Howells, R. (1987). *Farewell the Islands*. London: Gomer Press.

Igic, B., Fecheyer-Lippends, D., Xiao, M, Chan, A., Hanley, D., Brennan, P. R. L., Grim, T., Waterhouse, G. I. N., Hauber, M. E., and Shawkey, M. D. (2015). A nanostructural basis for gloss of avian eggshells. *Journal of the Royal Society Interface* 12: 20141210. http://dx.doi.org/10.1098/rsif.2014.1210

Ingold, P. (1980). Anpassungen der Eier und des Brutverhaltens von Trottellummen (*Uria aalge aalge* Pont.) an das Brüten auf Felssimsen. *Zeitschrift für Tierpsychologie* 53: 341–88.

Ishikawa, S., Suzuki, F., Fukuda, E., Arihara, K., Yamamoto, Y., Mukai, T., and Itoh, M. (2010). Photodynamic antimicrobial activity in avian eggshells. *FEBS Letters* 584: 770–5.

Jetz, W., Thomas, G. H., Joy, J. B., Hartmann, K., and Mooers, A. O. (2012). The global diversity of birds in space and time. *Nature* 491: 444–8.

Johnson, A. L. (2000). Reproduction in the female, pp. 569–96, in: *Sturkie's Avian Physiology* (ed. G. C. Whittow). San Diego: Academic Press.

Juniper, T. (2003). *Spix's Macaw: The Race to Save the World's Rarest Bird*. London: Fourth Estate/Atria.

Kaftanovski, Yu. M. (1941). Experiment in comparative characteristics of the biology of reproduction of several alcids. *Transactions of the Seven Islands Sanctuary* 1: 47–52. [In Russian and cited in Uspenski 1953]

Kaftanovski, Yu. M. (1951). Alcidine Birds (Alcids) of the Eastern Atlantic. *Materialy k poznaniyu fauny i flory SSSR*, Novaya Seria Otdel Zoologicheskii 28 (xiii) 10–169.

Kennedy, G. Y., and Vevers, H. G. (1976). A survey of avian eggshell pigments. *Comparative Biochemistry and Physiology* 55B: 117–23.

Kern, M. D., Cowie, R. J., and Yeager, M. (1992). Water loss, conductance, and structure of eggs of pied flycatchers during egg laying and incubation. *Physiological Zoology* 65: 1162–87.

Kerridge, R. 2014. *Cold Blood: Adventures with Reptiles and Amphibians*. London: Chatto & Windus.

Kightly, C. (1984). *Country Voices: Life and Lore in Farm and Village*. London: Thames & Hudson.

Kilner, R. M. (2006). The evolution of egg colour and patterning in birds. *Biological Reviews* 81: 383–406.

Kutter, F. (1877–8). Betrachtung über Systematik und Oologie vom Standpunkte der Selectionstheorie [Observations on systematics and oology from the point of view of the theory of natural selection], II. Teil. *Journal für Ornithologie* 25: 396–423; and 26: 300–48.

Lack, D. (1968). *Ecological Adaptations for Breeding in Birds*. London: Methuen.

Lang, M. R., and Wells, J. W. (1987). A review of eggshell pigmentation. *World's Poultry Science Journal* 43: 238–46.

Lipar, J. L., Ketterson, E. D., Nolan, V. Jr, and Casto, J. M. (1999). Egg yolk layers vary in the concentration of steroid hormones in two avian species. *General and Comparative Endocrinology* 115: 220–7.

Lorenz, K. (1965). *Evolution and Modification of Behaviour.* Chicago: University of Chicago Press.

Lovell, P. G., Ruxton, G. D., Langridge, K. V., and Spencer, K. A. (2013). Egg-laying substrate selection for optimal camouflage by quail. *Current Biology* 23: 260–4.

Lowther, P. E. (1988). Spotting pattern of the last laid egg of the house sparrow. *Journal of Field Ornithology* 59: 51–4.

Lyon, B. E. (2003). Egg recognition and counting reduce costs of avian conspecific brood parasitism. *Nature* 422: 495–9.

Lyon, B. E. (2007). Mechanism of egg recognition in defenses against conspecific brood parasitism: American coots (*Fulica americana*) know their own eggs. *Behavioral Ecology & Sociobiology* 61: 455–63.

McAldowie, A. M. (1886). Observations on the development and the decay of the pigment layer on birds' eggs. *Journal of Anatomy & Physiology* 20: 225–37.

McElwain, J. C., and Chaloner, W. G. (1996). The fossil cuticle as a skeletal record of environmental change. *Palaios* 11: 376–88.

MacGillivray, W. (1852). *A History of British Birds.* London: Scott, Webster & Geary.

MacLean, S. F. (1974). Lemming bones as a source of calcium for Arctic sandpipers (*Calidris* spp.). *Ibis* 116: 552–7.

McMaster, D. G., Sealy, S. G., Gill, S. A., and Neudorf, D. L. (1999). Timing of egg laying in yellow warblers. *Auk* 116: 236–40.

Magi, M., Mand, R., Konovalvov, A., Tilgar, V., and Reynolds, S. J. (2012). Testing the structural-function hypothesis of eggshell maculation in the great tit: an experimental approach. *Journal of Ornithology* 153: 645–52.

Manson-Bahr, P. (1959). Recollections of some famous British ornithologists. *Ibis* 101: 53–64.

Martin, M. (1698). *A Late Voyage to St Kilda, the Remotest of All the Hebrides, or Western Isles of Scotland.* London: Gent.

Maurer, G., Russell, G. D. and Cassey, P. (2010). Interpreting the lists and equations of egg dimensions in Schönwetter's *Handbuch der Oologie.* *The Auk*: 127: 940–947.

Maurer, G., Portugal, S. J., and Cassey, P. (2011). Review: An embryo's eye view of avian eggshell pigmentation. *Journal of Avian Biology* 42: 494–504.

Maurer, G., Portugal, S. J., Hauber, M. E., Miksik, I., Russell, D. G. D., and Cassey, P. (2015). First light for avian embryos: eggshell thickness and pigmentation mediate variation in development and UV exposure in wild bird eggs. *Functional Ecology* 29: 209–18.

Meckel, H. von Hemsback (1851). Die Bildung der für die partielle Furchung bestimmte Eier der Vögel im Vergleich mit den Graafschen Follikel und die Decidua des Menschen. *Zeitschrift für wissenschaftliche Zoologie* Bd 3: 420.

Montevecchi, W. A. (1976). Field experiments on the adaptive significance of avian eggshell pigmentation. *Behaviour* 58: 26–39.

Moreno, J., Morales, J., Lobato, E., Merino, S., Tomas, G., and Martinez-de la Puente, J. (2005). Evidence for the signalling function of egg colour in the pied flycatcher *Ficedula hypoleuca*. *Behavioral Ecology* 16: 931–7.

Moreno, J., and Osorno, J. L. (2003). Avian egg colour and sexual selection: does eggshell pigmentation reflect female condition and genetic quality? *Ecology Letters* 6: 803–6.

Moreno, J., Osorno, J. L., Morales, J., Merino, S., and Tomas, G. (2004). Egg colouration and male parental effort in the pied flycatcher *Ficedula hypoleuca*. *Journal of Avian Biology* 35: 300–4.

Morris, F. O. (1856). *A History of British Birds*. London: Groombridge.

Newton, A. (1896). *A Dictionary of Birds*. London: A. & C. Black.

Nice, M. M. (1937). Studies in the life history of the song sparrow. *Transactions of the Linnaean Society of New York* 4: 1–246.

Norton, D. W. (1972). Incubation schedules of four species of calidridine sandpipers at Barrow, Alaska. *Condor* 74: 164–76.

Nowak, E. (2005). *Wissenschaftler in turbulenten Zeiten*. Schwerin: Stock & Stein.

Nys, Y., and Guyot, N. (2011). Egg formation and chemistry, pp. 83–132, in: *Improving the Safety and Quality of Eggs and Egg Products*, vol. 1: *Egg Chemistry, Production and Consumption*. Oxford: Woodhead.

Opel, F. M. E. (1858). Beiträge zur Kenntniss des *Cuculus canorus* Lin. *Journal für Ornithologie* 6: 205–25.

Ord, G. (1836). Observations on the cow bunting of the United States of America. *Magazine of Natural History* 9: 57–71.

Oreskes, N., and Conway, E. M. (2010). *Merchants of Doubt*. New York: Bloomsbury.

Orians, G. H., and Janzen, D. H. (1974). Why are embryos so tasty? *American Naturalist* 108: 581–92.

Pappenheimer, J. (1996). *Hermann Rahn (1912–1990): A Biographical Memoir*. Washington: National Acadamies Press.

Parker, G. A. (1998). Sperm competition and the evolution of ejaculates: towards a theory base, pp. 3–54, in: *Sperm Competition and Sexual Selection* (eds T. R. Birkhead and A. P. Moller). London: Academic Press.

Payne, R. B. (1972). Nuts, bones, and a nesting of red crossbills in the Panamint Mountains. *Condor* 74: 485–6.

Pearl, R., and Curtis, M. R. (1912). Studies on the physiology of reproduction in the domestic fowl. VIII. On some physiological effects of ligation, section and removal of the oviduct. *Papers from the Biological Laboratory of the Maine Agricultural Experiment Station* no. 68: 395–424.

Pearl, R., and Schoppe, W. F. (1921). Studies on the physiology of reproduction in the domestic fowl. XVIII. Further observations on the anatomical basis of fecundity. *Journal of Experimental Zoology* 34: 101–18.

Pennant, T. (1768). *British Zoology*. London: Benjamin White.

Pike, T. W. (2015). Modelling eggshell maculation. *Avian Biology Research* 8: 237–243.

Pitmann, C. R. S. (1964): 'Eggs, natural history of', pp. 237–42, in: Thomson, A. L. (ed.) *A New Dictionary of Birds*. London: Nelson.

Priestley, J. B. (1921). *I, for One*. Oxford: Bodley Head.

Prynne, M. (1963). *Egg-shells*. London: Barrie & Rockliff.

Purcell, R., Hall, L. S., and Coardso, R. (2008). *Egg & Nest*. Cambridge, Mass.: Belknap Press.

Purkinje, J. E. (1930). *Symbolic ad ovi avium historiam ante incubationem*. Leipzig: Vossius.

Pycraft, W. P. (1910). *A History of Birds*. London: Methuen.

Rahman, M. A., Baoyindeligeer, Iwasawa A., and Yoshizaki, N. (2007). Mechanism of chalaza formation in quail eggs. *Cell Tissue Research* 330: 535–43.

Rahn, H. (1991). Why birds lay eggs, pp. 345–60, in *Egg Incubation: Its Effects on Embryonic Development in Birds and Reptiles* (eds C. Deeming and M. W. J. Ferguson). Cambridge: Cambridge University Press.

Rahn, H., Ar, A., and Paganelli, C. v. (1979). How eggs breathe. *Scientific American* 240: 38–47.

Rahn, H., Carey, C., Balmas, K., Bhatia, B., and Paganelli, C. V. (1977). Reduction in pore area of the avian eggshell as an adaptation. *Proceedings of the National Academy of Sciences of the USA* 74: 3095–8.

Rahn, H., Ledoux, T., Paganelli, C. V. and Smith, A. H. (1982). Changes in eggshell conductance after transfer of hens from an altitude of 3,800 to 1,200m. *Journal of Applied Physiology: Respiratory, Environmental and Exercise Physiology.* 53: 1429–1431.

Rahn, H., and Paganelli, C. V. (1991). Energy budget and gas exchange of avian eggs, pp. 175–93, in: *Avian Incubation* (ed. S. G.Tullett). London: Butterworth & Heinemann.

Rahn, H., Paganelli, C. V., and Ar, A. (1987). Pores and gas exchange in avian eggs: a review. *Journal of Experimental Zoology*, Supplement 1: 165–72.

Ray, J. (1678). *The Ornithology of Francis Willughby*. London: John Martyn.

Ray, J. (1691). *The Wisdom of God Manifested in the Works of Creation*. London: Smith.

Réaumur, M. de (1750). *The Art of Hatching and Bringing up Domestic Fowls of all Kinds, at any time of year, either by means of hot-beds, or that of common fire*. Paris: Royal Academy of Sciences.

Rensch, B. (1947). *Neuere Probleme der Abstammungslehre*. Stuttgart: Ferdinand Enke.

Reynolds, J., and Perrins, C. M. (2010). Dietary calcium availability and reproduction in birds. *Current Ornithology* 17: 31–74.

Reynolds, S. J., Martin, G. R., and Cassey, P. (2009). Is sexual selection blurring the functional significance of eggshell coloration hypotheses? *Animal Behaviour* 78: 209–15.

Riehl, C. (2010). A simple rule reduces costs of extragroup parasitism in a communally breeding bird. *Current Biology* 29: 1830–3.

Rogers, C. A. (1908). Feeding color – an aid in studying physiological development. *Poultry Science* S1: 76–81.

Romanoff, A. J., and Romanoff, A. L. (1949). *The Avian Egg*. New York: Wiley.

Roudybush, T. E., Grau, C. R., Petersen, M. R., Ainley, D. G., Hirsch, K. V., Gilsan, A. P., and Patten, S. M. (1979). Yolk formation in some charadriiform birds. *Condor* 81: 293–8.

Russell, G. D., et al. (2010). Data-poor egg collections: cracking an important research resource. *Journal of Afrotropical Zoology* Special Issue 6: 77–82.

Salamon, A., and Kent, J. P. (2014). Orientation of the egg at laying – is the pointed or blunt end first? *International Journal of Poultry Science* 13: 316–18.

Salmon, M. A. (2000). *Aurelian Legacy: A History of British Butterflies and their Collectors*. Leiden: Brill.

Sauveur B., and Mongin, P. (1974). Effects of time limited calcium meal upon food and calcium ingestion and egg quality. *Br. Poult Sci.* 15:305–313.

Schmidt, W. J. (1956). Beiträge zur mikroskopischen Kenntnis der Farbstoffe in der Kalkschale des Vogeleies. *Zeitschrift für Zellforschung* 44: 413–26.

Schulze-Hagen, K. and Birkhead, T. R. (2015). The ethology and life history of birds: the forgotten contributions of Oskar, Magdalena and Katharina Heinroth. *Journal of Ornithology*. 156: 9–18.

Schulze-Hagen, K., Stokke, B., and Birkhead, T. R. (2009). Reproductive biology of the European Cuckoo *Cuculus canorus*: early insights, persistent errors and the acquisition of knowledge. *Journal of Ornithology* 150: 1–16.

Schwabl, H. 1993. Yolk is a source of maternal testosterone for developing birds. *Proceedings of the National Academy of Sciences of the USA* 90: 11466–11450.

Seebohm, H. (1883). A *History of British Birds, with Coloured Illustrations of their Eggs*. London: R. H. Porter.

Short, R. (2003). The magic and mystery of the oocyte: Ex ovo Omnia, pp. 3–10, in: *Biology and Pathology of the Oocyte* (eds A. O. Trounson and R. G. Gosden). Cambridge: Cambridge University Press.

Siegfried, R. (2008). John Davy in: *The Complete Dictionary of Scientific Biography*. http://www.encyclopedia.com/doc/1G2–2830901095.html

Sinervo, B. (1993). The effect of offspring size on physiology and life history. *Bioscience* 43: 210–18.

Snook, R. R., Hosken, D. J., and Karr, T. L. (2011). The biology and evolution of polyspermy: insights from cellular and functional studies of sperm and centrosomal behavior in the fertilized egg. *Reproduction* 142: 779–92.

Soler, J. J., Martin-Vivaldi, M., Ruiz-Rodrigues, M., Valdivia, E., Martin-Platero, A. M., Martinez-Bueno, M., Peralta-Sanchez, J. M., and Mendez, M. (2008). Symbiotic association between hoopoes and antibiotic-producing bacteria that live in the uropygial gland. *Functional Ecology* 22: 864–71.

Sorby, H. C. (1875). On the colouring-matters of the shells of birds' eggs. *Proceedings of the Zoological Society of London* 23: 351–65.

Sorensen, M. D., and Payne, R. B. (2001). A single ancient origin of brood parasitism in African finches: implications for host-parasite evolution. *Evolution* 55: 2550–67.

Sossinka, R. (1982). Domestication in birds. *Avian Biology* 7: 373–403 (eds D. S. Farner, J. R. King and K. C. Parkes). New York: Academic Press.

Sotherland, R. R. and Rahn, H. (1987). On the composition of bird eggs. *Condor* 89: 48–65.

Sotherland, P. R., Wilson, J. A., and Carney, K. M. (1990). Naturally occurring allometric engineering experiments in avian eggs. *American Zoologist* 30: 86A.

Sparks, N. H. C. (1994). Shell accessory materials: structure and function, pp. 25–42, in: *Microbiology of the Avian Egg* (eds R. G. Board and R. Fuller). London: Chapman & Hall.

Stepinska, U., and Bakst, M. R. (2007). Fertilization, pp. 553–87, in: *Reproductive Biology and Phylogeny of Birds* (ed. B. J. Jamieson). Enfield: Science Publishers.

Strassmann, B. I. (2013). Concealed ovulation in humans: further evidence, pp. 139–51, in: *Human Social Evolution* (eds K. Summers and B. Crespi). Oxford: Oxford University Press.

Stresemann, E. (1975). *Ornithology from Aristotle to the Present*. Harvard: Harvard University Press.

Swynnerton, C. F. M. (1916). On the coloration of the mouths and eggs of birds. II. On the coloration of eggs. *Ibis* 4: 529–606.

Sykes, A. H. (1953). Some observations on oviposition in the fowl. *Quarterly Journal of Experimental Physiology* 38: 61–8.

Tamura, T., and Fuji, S. (1966). Histological observations on the quail oviduct: histochemical observations on the secretions of the glands and the mucous cells. *Journal of the Faculty of Fish & Animal Husbandry Hiroshima University* 6: 373–93.

Tarchanoff, J. R. (1884). Uber die Verschiedenheiten des Eierweisses bei gefiedert geborenen (Nestflüchter) und bei nackt geborenen (Nesthocker) Vögeln. *Pflügers Archiv für die gesamte Physiologie des Menschen und der Tiere* 33: 363–78.

Taschenberg, O. (1885). Zur frage über die Entstehung der Farbung der Vogeleierschalen. *Zoologischer Anzeiger* 8: 243–5.

Thaler, E. (1990). *Die Goldhähnchen. Die Neue Brehm Bücherei*. Wittenberg: Ziemsen.

Thienemann, F. A. L. (1843). *Systematische Darstellung der Fortpflanzung der Vögel Europas mit Abbildung der Eier* [A Systematic Account of the breeding of European birds with illustrations of their eggs]. Lepizig: I. A. Barth.

Thompson, D'A. W. (1908). Shapes of eggs. *Nature* 78: 111–13.

Thompson, D'A. W. (1917). *On Growth and Form.* Cambridge: Cambridge University Press.

Thomson, J. A. (1923). *The Biology of Birds.* New York: Macmillan.

Thomson. A. L. (1964). *A New Dictionary of Birds.* London: Nelson.

Tiedemann, F. (1810, 1814). *Anatomie und Naturgeschichte der Vögel.* Heidelberg: Mohr & Zimmer.

Tilly, J. L., Niikura, Y., and Rueda, B. R. (2009). The current status of evidence for and against postnatal oogenesis in mammals: a case of ovarian optimism versus pessimism? *Biology of Reproduction* 80: 2–12.

Tinbergen, N., Broekhuysen, G. J., Feekes, F., Houghton, J. C., Kruuk, H., and Szulc, E. (1962). Egg shell removal by the black-headed gull: a behaviour component of camouflage. *Animal Behaviour* 19, 74–117.

Topsell, E. (1625). *The Fowles of Heauen or History of Birds.* Austin, Tex: University of Texas (1972).

Tordoff, M. G. (2001). Calcium: taste, intake, and appetite. *Physiological Reviews* 81: 1567–97.

Tschanz, B. (1968). Trottellummen: Die Entstehung der personlichen Beziehung zwischen Jungvogel und Eltern. *Zeitschrift für Tierpsychologie* Biheft 4.

Tschanz, B. (1990). Adaptations for breeding in Atlantic alcids. *Netherlands Journal of Zoology* 40: 688–710.

Tschanz, B., Ingold, P., and Lengacher, H. (1969). Eiform und Bruterfolg bei Trottellummen (*Uria aalge*). *Ornithologische Beobachter* 66: 25–42.

Tuck, L. M. (1961). *The Murres: Their Distribution, Populations and Biology, as study of the genus Uria.* Ottawa: Canadian Wildlife Service.

Tyler, C. (1964). *Wilhelm von Nathusius 1821–1899 on Avian Eggshells.* Reading: Berkshire Printing.

Tyler, C. (1969). Avian eggshells: Their structure and characteristics. *International Review of General and Experimental Zoology* 4: 82–127.

Uspenski, S. M. (1958). *The Bird Bazaars of Novaya Zemlya*. Ottawa: Department of Northern Affairs and National Resources, Canada.

Vaughan, R. (1998). *Seabird City: A Guide to the Breeding Seabirds of the Flamborough Headland*. Otley: Smith Settle.

Vince, M. A. (1969). Embryonic communication, respiration and synchronisation of hatching. Chapter 11 in: *Bird Vocalisations* (ed. Hinde, R. A.). Cambridge: Cambridge University Press.

Vincze, O., Vagasi, C. I., Kovacs, I., Galvan, I., and Pap, P. L. (2013). Sources of variation in uropygial gland size in European birds. *Biological Journal of the Linnean Society* 110: 543–63.

Waldeyer, W. (1870). *Eierstock und Ei*. [Ovary and Egg]. Leipzig: Engelmann.

Wallace, A. R. (1889). *Darwinism: An Exposition of the Theory of Natural Selection, with some of its Applications*. London: Macmillan.

Wallace, A. R. (1895). *Natural Selection and Tropical Nature*. New York: Macmillan & Co.

Wallace, W. H. B., and Kelsey, T. W. (2010). Human ovarian reserve from conception to the menopause. *PLoS ONE* 5(1): e8772. doi:10.1371/journal.pone.0008772

Walters, M. P. (1994). *Birds' Eggs*. London: Dorling Kindersley.

Warham, J. (1990). *The Petrels: Their Ecology and Breeding Systems*. London: Academic Press.

Waterton, C. (1835). Notes of a visit to the haunts of the guillemot and facts on its habits. *Magazine of Natural History and Journal of Zoology, Botany, Mineralogy, Geology and Meteorology* 8: 162–5.

Waterton, C. (1871). *Essays on Natural History, Chiefly Ornithology: with an Autobiography of the Author and a view of Walton Hall*. London: Mawman.

Webster, R. J., Callahan, A., Godin, J-G. J., and Sherratt, T. N. (2009). Behaviourally mediated crypsis in two nocturnal moths with contrasting appearance. *Philosophical Transactions of the Royal Society B: Biological Sciences* 364: 503–10.

Whittaker, J. (ed.) (1997). *A Diary of Bempton Climbers*. Leeds: Peregrine Books.

Whitteridge, G. (1981). *Disputations Touching the Generation of Animals*. Oxford: Blackwell.

Wicke, W. (1858). Ueber das pigment in den Eischalen der Vogel. *Naumannia* 8: 393–7.

Wickmann, H. (1896). Die Lage des Vogeleies vor und während der Geburt. *Journal für Ornithologie* 44: 81–92.

Wiemann, J., Yang, T-R., Sander, P. N., Schneider, M., Engeser, M., Kath-Schorr, S., Müller, C. E., and Sander, P. M. (2015). The blue-green eggs of dinosaurs: how fossil metabolites provide insights into the evolution of bird reproduction. *Peer J Preprint*. 3:e1323 doi: https://dx.doi.org/10.7287/peerj.preprints.1080v1 https://peerj.com/preprints/1080v1/

Williams, T. D. (1995). *The Penguins*. Oxford: Oxford University Press.

Wirsling, A. L., and Günther, F. C. (1772). *Sammlung von Nestern und Eyern verschiederener Vogel*.

Wood, C. A. (ed.) (1958). *The Continuation of the History of the Willughby Family by Cassandra Duchess of Chandos*. Windsor: University of Nottingham.

Yarrell, W. (1826). On the small horny appendage to the upper mandible in very young chickens. *Zoological Journal* 1826: 433–7.

Yarrell, W. (1843). *A History of British Birds*. London: Van Voorst.

Zuckerman, S. (1965). The natural history of an enquiry. *Annals of the Royal College of Surgeons of England* 37: 133–49.

Glossary

Altricial chicks   Fairly helpless, naked and unable to see (eyes closed) when they hatch (cf. precocial chicks).

Arms races   Occur when two actors (individuals, groups or species) have a conflict of interests. As a result of this conflict, one actor changes its behaviour, morphology or physiology to get the upper hand and the other responds to reduce the cost to itself. In a co-evolutionary arms race, an adaptation in one actor results in the evolution of a counter-adaptation in the other actor, and this can occur both within and between species. The adaptations to brood parasitism (e.g. egg mimicry) by brood parasites, and the counter-adaptations (egg recognition) by the host is an example, and can occur within and between species.

Brood parasite   A bird that lays its eggs in the nest of another bird. Brood parasites can be either interspecific – laying in the nest of another species, like the familiar common cuckoo – or intraspecific – parasitising members of its own species, like the common starling and the American coot.

Brood patch   A region of featherless skin – there may one, two or three depending on the species – on a bird's abdomen through which heat is transmitted to the egg(s).

Chalazae   The two twisted strands of albumen attached to opposite sides of the yolk that keep it in position within the egg and also allow it to rotate, such that the embryo is kept uppermost.

Chorio-allantois   A vascularised structure in birds' eggs formed by the fusion of two membranes – the chorion and allantois – creating a network of fine blood vessels lying against the shell membrane and connected to the embryo. It fulfils the same role as the placenta in mammals.

Cloaca Sometimes called the vent in birds, this is the common posterior opening of the alimentary canal (anus), the urinary tract (ureter) and the reproductive system (vagina in females; vas deferens in male birds).

Cuticle The thin, outermost organic covering on an eggshell (see also shell accessory material).

Egg tooth A hard usually white protuberance on the tip of the upper – and sometimes the lower – mandible in fully developed embryos (chicks), used to break out of the egg.

Functional explanation One that considers the adaptive significance of a biological phenomenon. Functional questions or explanations address 'why' questions (cf. Mechanistic explanations): why did this feature evolve, why does it enhance survival or reproductive success?

Generation An old term that refers to the combined processes of sexual reproduction and embryo development.

Germinal disc (or blastodisc) That part of an unfertilised ovum that contains the female genetic material. It is usually a pale spot on the surface of the yolk and is where fertilisation takes place.

Gizzard The bird's muscular stomach. It is not to be confused with the bird's crop, a temporary storage container between the mouth and the stomach.

Incubation period The interval between the start of incubation and hatching. It is not necessarily the same as the interval between the start of laying and hatching because some birds delay the start of incubation after the first egg is laid.

Infundibulum The most anterior (closest to the head) region of the oviduct, immediately adjacent to the ovary. This is where fertilisation occurs.

Maculation The markings on eggs.

Magnum The region of the oviduct between the infundibulum and isthmus.

Mechanistic explanation One that addresses the 'how?' questions of a biological phenomenon (cf. Functional explanation); how does the physiology or mechanics of the organism influence the trait of interest?

Oocyte The immature female sex (or reproductive) cell in the mammalian ovary, from which an ovum develops.

Ovary The female reproductive organ that produces the ova. In humans and other mammals the ovaries are paired; in virtually all birds there is only a single (left) ovary.

Oviduct The tube through which the ovum and egg passes from the ovary to the outside world becoming a shelled egg in the process. The oviduct comprises several different regions (e.g. Infundibulum, Magnum, etc.).

Oviposition The fancy term for egg laying. Eggs are laid or oviposited.

Ovulation The release of the unfertilised ovum (or oocyte) into the oviduct from the ovary.

Ovum The mature female sex cell (plural, ova).

Passerine A perching bird or songbird; more than half of all birds are passerines (cf. Non-passerines).

pH A logarithmic scale – from 1 to 14 – on which acidity (low values) and alkalinity (high values) is measured.

Pipping The act of breaking through the eggshell by the chick just before it hatches.

Polyspermy Literally means 'many sperm'. Pathological polyspermy, which can occur in mammals, is the accidental result of several (rather than just one) sperm penetrating the unfertilised ovum and causing embryo death. Physiological polyspermy, which occurs in birds, sharks and some amphibians, is a normal part of the fertilisation process.

Porphyrin or protoporphyrin A dark, reddish pigment that forms the maculation on birds' eggs.

Precocial chicks   These hatch with their eyes open, and are able to walk, to feed themselves, and, in the case of megapodes, able to fly (cf. Altricial chicks).

Pyriform   One shape of an egg: rounded at one end and pointed at the other.

Runt egg or dwarf egg   One that is smaller than is typical for the species and usually contains no yolk and hence cannot hatch.

Shell membrane   A double, very thin fibrous layer surrounding the albumen, and in contact with the inner surface of the eggshell.

Shell accessory material   The outermost covering of an eggshell. The term encompasses both the organic (cuticle) and inorganic (calcium salts) layers on the outside of eggshells.

Sperm storage tubules   Tiny, blind-ending sausage-shaped tubules located in the oviduct at the junction between the vagina and the uterus where sperm are stored prior to their being transported to the infundibulum for fertilisation.

Superhydrophobic surface   One that repels water and has a 'water contact angle' great than 150°. Any hydrophobic surface repels water to some extent.

Uterus   A region of the oviduct between the isthmus and the vagina, also called the shell gland. The bird's uterus comprises a 'red-region' and, more posteriorly, a bag-like region.

Vagina   The region of the oviduct furthest from the ovary, and closest to the cloaca adjacent to the uterus.

Vaterite   A form of calcium carbonate.

Yolk sac   A membrane that encloses the yolk in a developing bird's egg.

Bird Species Mentioned in the Text

These are the species mentioned by name in the text. I have used the name here as it appears in the text, but I also provide the full name recognised by the International Ornithological Committee (www.worldbirdnames.org). Groups of birds, such as 'shearwaters' or 'woodpeckers' mentioned in the text are not listed here. Species are arranged here in alphabetical order for convenience.

African thrush *Turdus pelios*
Alpine swift *Tachymarptis melba*
American coot *Fulica americana*
American rhea *Rhea americana*
American robin *Turdus migratorius*
Argus (great) pheasant *Argusianus argus*
Australian brushturkey *Alectura lathami*
Avocet (Pied) *Recurvirostra avocetta*
Barn owl *Tyto alba*
Barn swallow *Hirundo rustica*
Black-browed albatross *Thalassarche chrysostoma*
Black guillemot *Cepphus grylle*
Black-headed gull *Larus ridibundus*
Blackbird *Turdus merula*
Blue tit *Cyanistes caerules*
Bobwhite quail *Colinus virginianus*
Brown-headed cowbird *Molothrus ater*
Brünnich's guillemot *Uria lomvia*

Budgerigar *Melopsittacus undulatus*
Buzzard (Common) *Buteo buteo*
California condor *Gymnogyps californianus*
Canary (domesticated) *Serinus canaria*
Canvasback *Aythya valisneria*
Caspian tern *Hydroprogne caspia*
Cassin's auklet *Ptychoramphus aleuticus*
Chaffinch (Common) *Fringilla coelebs*
Common eider *Somateria mollissima*
Common guillemot *Uria aalge*
Common kestrel *Falco tinnunculus*
Common sandpiper *Actitis hypoleucus*
Common starling *Sturnus vulgaris*
Corn bunting *Emberiza calandra*
Crane (Common) *Grus grus*
Crossbill (Common) *Loxia curvirostra*
Crow (Carrion) *Corvus corvus*
Crowned sandgrouse *Pterocles coronatus*
Cuckoo (Common) *Cuculus canorous*
Cuckoo finch *Anomalospiza imberbis*
Curlew (Eurasian) *Numenius arquata*
Domestic fowl (chicken) *Gallus domesticus*
Double-banded courser *Rhinoptilus africanus*
Dunnock *Prunella modularis*
Egyptian plover *Pluvianus aegyptius*
Egyptian vulture *Neophron percnopterus*
Elephant bird *Aepyornis maximus*
Emperor penguin *Aptenodytes forsteri*
Emu *Dromaius novahollandiae*
Eurasian dipper *Cinclus cinclus*
Eurasian sparrowhawk *Accipiter nisus*
European roller *Coracias garrulus*
Fiordland crested penguin *Eudyptes pachyrhynchus*

Firecrest *Regulus ignicapilla*
Fork-tailed storm petrel *Oceanodroma furcata*
Gannet (Northern) *Morus bassana*
Garden warbler *Sylvia borin*
Glaucous gull *Larus hyperboreus*
Goldcrest *Regulus regulus*
Golden plover (Eurasian) *Pluvialis apricaria*
Goldfinch *Carduelis carduelis*
Great auk *Alca impennis*
Great black-backed gull *Larus marinus*
Great cormorant *Phalacrocorax carbo*
Great northern diver (Common loon) *Gavia immer*
Great tinamou *Tinamus major*
Great tit *Parus major*
Greater ani *Crotophaga major*
Greater bowerbird *Chlamydera nuchalis*
Green woodpecker *Picus viridis*
Grey gull *Leucophaeus modestus*
Grey-faced petrel *Pterodroma macroptera gouldi*
Grey wagtail *Motacilla cinerea*
Griffon vulture *Gyps fulvus*
Guira cuckoo *Guira guira*
Hen Harrier (Northern harrier) *Circus cyaneus*
Heron (grey) *Ardea cinerea*
Herring gull *Larus argentatus*
Hihi (Stitchbird) *Notiomystis cincta*
Hoopoe *Upupa epops*
House sparrow *Passer domesticus*
Japanese quail *Coturnix coturnix japonica*
King penguin *Aptenodytes patagonicus*
Kittiwake (black-legged) *Rissa tridactyla*
Kiwi (North Island Brown) *Apteryx mantelli*
Lapwing (Northern) *Vanellus vanellus*

Laughing gull *Leucophaeus atricilla*
Lesser black-backed gull *Larus fuscus*
Lesser whitethroat *Sylvia curruca*
Little bittern *Ixobrychus minutus*
Little tern *Sternula albifrons*
Long-tailed tit *Aegithalos caudatus*
Magpie (Eurasian) *Pica pica*
Mallard *Anas platyrhynchos*
Malleefowl *Leipoa ocellata*
Manx shearwater *Puffinus puffinus*
Mistle thrush *Turdus viscivorus*
Nightingale *Luscinia megarhynchos*
Northern fulmar *Fulmarus glacialis*
Osprey *Pandion haliaetus*
Ostrich *Struthio camelus*
Oystercatcher (Eurasian) *Haematopus ostralegus*
Palm (African) swift *Cypsiurus parvus*
Peacock *Pavo cristatus*
Pearly-eyed thrasher *Margarops fuscatus*
Peregrine falcon *Falco peregrinus*
Philippine megapode *Megapodius cumingii*
Pied flycatcher *Ficedula hypoleuca*
Pigeon (domestic) *Columba livia*
Puffin (Atlantic) *Fratercula arctica*
Raven (Common) *Corvus corax*
Razorbill *Alca torda*
Red junglefowl *Gallus gallus*
Red phalarope (grey phalarope in Europe) *Phalaropus fulicarius*
Red-cockaded woodpecker *Picoides borealis*
Red-faced cisticola *Cisticola erythrops*
Ring ouzel *Turdus torquata*
Ring-necked parakeet *Psittacula krameri*
Ringed (Common) plover *Charadrius hiaticula*

Robin (Eurasian) *Erithacus rubeluca*
Rock ptarmigan *Lagopus muta*
Rook *Corvus frugilegus*
Ross's turaco *Musophaga rossae*
Royal tern *Thalasseus maximus*
Sandwich tern *Thalasseus sandvicensis*
Scops (Eurasian) owl *Otus scops*
Short-tailed shearwater *Ardenna tenuirostris*
Shoveller *Anas clypeata*
Sky lark (Eurasian) *Alauda arvensis*
Slavonian grebe *Podiceps auritus*
Sociable weaver *Philetarius socius*
Song thrush *Turdus philomelus*
Southern brown kiwi *Apteryx australis*
Southern royal albatross *Diomedea epomorphora*
Spix's macaw *Cyanopsitta spixii*
Spotted bowerbird *Chlamydera maculata*
Storm petrel (European) *Hydrobates pelagicus*
Striated heron *Butorides striata*
Tawny-flanked prinia *Prinia subflava*
Tufted duck *Aythya fuligula*
Wattled jacana *Jacana jacana*
Whimbrel *Numenius phaeopus*
White stork *Ciconia ciconia*
Willow ptarmigan *Lagopus lagopus*
Woodpigeon *Columba palumbus*
Wren (Eurasian) *Troglodytes troglodytes*
Wryneck *Jynx torquilla*
Yellow warbler *Setophaga petechial*
Yellow-breasted bowerbird *Chlamydera lauterbachi*
Yellowhammer *Emberiza citrinella*
Zebra finch *Taeniopygia guttata*

Acknowledgements

In 2003 I was given a copy of William Hewitson's beautifully illustrated *British oology*, published in 1831. It was a gift from a community of reproductive biologists in gratitude for the biennial scientific meeting on eggs and sperm I have organised (with my colleague Harry Moore) since 1992. Scott Pitnick, who has attended these meetings since the start, asked the other delegates to contribute to a present and, without my knowing, asked my wife Miriam what old bird books I did *not* have. In due course he found and purchased Hewitson's two volumes. At the meeting I was taken completely by surprise when Scott presented me with the books. Nor did I realise at the time that they set the seed for this volume.

During the research and writing I have received some wonderful help from a wide range of people, but I am especially grateful to the librarians or people who know librarians that tracked down obscure references for me: Chris Everest (University of Sheffield Library), Fiona Fisken (Zoological Society of London [ZSL]), Effie Warr (British Museum, Natural History [BMNH]), Linda da Volls (ZSL), John Simpson (Accrington Library), Ann Sylph (ZSL) and Mike Wilson (Alexander Library, Oxford).

The curators of several museums gave up their time to provide me with access to their egg collections. I am extremely grateful to Rob Barrett (Tromso), Julian Carter (Cardiff), Clem Fisher (Liverpool), Jan Fjeldsa (Copenhagen), Dan Gordon (Newcastle),

Henry McGhie (Manchester), Robert Prys-Jones and especially Douglas Russell (both at BMNH, Tring).

I also wish to thank the community of egg researchers for their help and willingness to discuss ideas: Phil Cassey, Charles Deeming, Mark Hauber, Steve Portugal and Jim Reynolds. Many of the findings discussed in this book are based on their excellent work.

Eleanor Caves, Bruce Lyon, Verity Peterson, Claire Spottiswoode and Chris Wallbank generously provided some of the images. David Quinn created some superb illustrations and Emily Glendenning drew the figures: I thank them both. Thanks also to Chris Wallbank whose extraordinary guillemot frieze we have included as the endpapers. I am extremely grateful to Sir David Attenborough for allowing me to examine his egg of an elephant bird *Aepyornis* and my brother Mike for the photograph of it that appears on the back jacket.

Other people who patiently answered my queries include Carry Akroyd, Kraig Adler, André Ancel, Patricia Brekke, Patty Brennan, Jean Pierre Brillard, Isabelle Charmantier, Nick Davies, Jim Flegg, Mark Geoghegan, Jeremy Greenwood, Alan Gilbert, Bill Hale, Len Hill, Paul Hocking, Paul Ingold, Yves Nys, Peter Lack, Toby Lupton, Mike McCarthy, Rebecca Manley, Peter Marren, Michael Middleton, Ian Newton, Natalie and Yuri Nikolaeva, Brian Oliver, Verity Peterson, Anna-Marie Roos, Richard Serjeantson, Stuart Sharp, Claire Spottiswoode, Hallvard Strøm, Craig Sturrock, Ellen Thaler, Jim Whitaker and Bernie Zonfrillo – I am grateful to them all. Karl Schulze-Hagen deserves a special mention: not only did he translate a lot of German ornithological literature for me, he helped in many other ways as well, for which I owe him immense thanks.

It is a pleasure to acknowledge my academic colleagues at Sheffield who have helped with my egg research in various different ways. These are: Keith Burnett, Ash Cadby, Caroline Evans, Patrick Fairclough, James Grinham, Nicola Hemmings, Duncan Jackson,

Roger Lewis, Tony Ryan, Vanessa Toulmin and Philip Wright. I am especially grateful to John Biggins, professor of statistics, who has been a good friend and a constant source of inspiration during our entire careers at Sheffield. I owe Jamie Thompson special thanks, too: it was he who accompanied me to various European museums and didn't baulk when, on Skomer Island, Wales, I asked him to crawl on to guano-covered guillemot ledges with me to undertake some of the experiments described in this book. I am also grateful to my numerous field assistants and the wardens in Skomer, past and present, for their help, and to the Wildlife Trust for South and West Wales for permission to work in one of the most beautiful places in the world.

I thank my friend John Barlow for his guillemot haiku:

all the colours
in the swell below . . .
guillemot eggs

Duncan Jackson, Bob Montgomerie and Jeremy Mynott generously gave up their time to read and comment on the entire manuscript. Their critical and helpful suggestions, honesty and friendship have been inspirational.

My agent Felicity Bryan, and my brilliant editor Bill Swainson at Bloomsbury and his team, and in particular Nick Humphrey, all provided wonderful, enthusiastic and pragmatic support for which I am extremely grateful.

For everything else I thank my wife Miriam and children, Nick, Fran and Laurie.

A Note on the Plate Section

The final colour plate is a photograph of eggs of different Zambian warblers and weaver birds parasitised by two brood parasites, the diederik cuckoo *Chrysococcyx caprius* and the cuckoo finch *Anomalospiza imberbis*. The host species have evolved signature-like variation in their egg colours and patterns both within and between species to help them to detect eggs of parasites that mimic their own (see text).

Each column shows a different host species: from left to right, red-faced cisticola (*Cisticola erythrops*), red-billed buffalo-weaver (*Bubalornis niger*), fan-tailed widowbird (*Euplectes axillaris*), white-winged widowbird (*Euplectes albonotatus*), Stierling's wren-warbler (*Calamonastes stierlingi*), grey-backed camaroptera (*Camaroptera brevicaudata*), desert cisticola (*Cisticola aridula*), southern masked weaver (*Ploceus velatus*), Holub's golden weaver (*Ploceus xanthops*), tawny-flanked prinia (*Prinia subflava*), pale-crowned cisticola (*Cisticola cinnamomeus*), rattling cisticola (*Cisticola chiniana*), yellow-crowned bishop (*Euplectes afer*), and white-browed sparrow-weaver (*Plocepasser mahali*). Photographs by Eleanor Caves and Claire Spottiswoode.

Index